A TIME OF TERROR

BOOKS BY J. BOWYER BELL

Besieged: Seven Cities under Attack

The Long War: Israel and the Arabs Since 1946

The Secret Army: The I.R.A., 1916–1974

The Myth of the Guerrilla:
Revolutionary Theory and Malpractice

The Horn of Africa: Strategic Magnet in the Seventies

Transnational Terror

On Revolt: Strategies of National Liberation

Terror Out of Zion: The Irgun Zvai Leumi, LEHI, and the
Palestine Underground, 1929–1949

A TIME OF TERROR

HOW DEMOCRATIC SOCIETIES RESPOND TO REVOLUTIONARY VIOLENCE

J. BOWYER BELL

BASIC BOOKS, INC., PUBLISHERS
NEW YORK

This book was written under the auspicies of the Institute of War and Peace Studies, Columbia University.

Library of Congress Cataloging in Publication Data

Bell, J. Bowyer, 1931–
A time of terror.

"Written under the auspicies of the Institute of War and Peace Studies, Columbia University."
Bibliography: p. 280
Includes index.
1. Terrorism. I. Columbia University. Institute of War and Peace Studies. II. Title.
HV6431.B43 301.6'33 77-20410
ISBN: 0-465-08621-7

Printed in the United States of America
DESIGNED BY VINCENT TORRE
10 9 8 7 6 5 4 3 2 1

IN MEMORIAM

JOHN MELVILLE MILLER, III

A gentleman, scholar, a physician
and musician, godfather, gourmet and wit,
a friend and reader I could
ill afford to lose.

. . . there are no innocent people.
Emile Henry

CONTENTS

PART THREE

THE RESPONSE: STRATEGIES OF THE THREATENED

PART FOUR

CONCLUSION

PREFACE

We live in a time of terror. Innocents are slaughtered, statesmen murdered, airplanes hijacked, and officials kidnapped by men and women who have brought terror out of the history books or the guerrilla wars of the bush and into the open, democratic societies of the postindustrial world. States without nationality problems find their capitals used as arenas for media spectaculars choreographed by strange men pursuing alien causes—a Free Croatia, a Liberated Palestine, or some other luminous vision of a perfect society. Acting out of loyalty to a higher cause, the new gunmen appear to be beyond compromise; even, at times, beyond reason. The threatened nations have reacted in varying ways, in response to differing traditions, predilections, and possibilities, and more often than not in isolation.

The result has been that there is no common wisdom, no consensus on an appropriate democratic response, often not even agreement on a national policy. Yet the fact remains that *something* must be done—to stop the hijacking, or rescue the hostages, or end sanctuary. And during the last few years a lot has been done, not all desirable, it is true, but not all futile either. Nowhere in the growing literature on terrorism, however, does there exist a rigorous, crossnational analysis of what the response of the threatened has been.

Clearly, the time has come for a general book on the liberal, democratic response to terrorism. For me such a project has two special fascinations. Most of my academic life has been spent on the other side of the barricades, investigating the guerrillas, assassins, and gunmen—indeed, some of my best friends are or have been "terrorists" in the public eye.

To look at what the threatened did would be a novel experience—especially when the threatened were the most vulnerable and valuable institutions of open, democratic societies. Then, also, terrorism as generally perceived seems to bring out the worst in most institutions and agencies. The very word becomes a litmus test for dearly held beliefs, so that a brief conversation on terrorist matters with almost anyone reveals a special world view, an interpretation of the nature of man, and a glimpse into a desired future. Some of those who write on terrorism seem to notice this, but many, including my academic colleagues, often do not, oblivious to the fact that they are shaping analysis into advocacy. At worst some of their proposals seem likely to close down open societies in order to make them safe, while at best they offer tactics that seem to work, but they fail to provide any convincing proof. While, given the obstacles to disinterested analysis, I might be able to offer little additional "proof," the opportunity remains to reveal what has or has not been accomplished, and at what cost, rather than, as has so often been done in the past, focusing on what should be done and the advocacy of favored prescriptions.

In one form or another I have been investigating revolutionary violence for two decades. I have spent years talking to members of the IRA and to Palestinian fedayeen, written books on Zionist terrorists and Cypriot gunmen, wandered in the wild places of Eritrea and Rhodesia, visited the garden spots of violence in Aden and Beirut and Belfast. The experience of these ventures—as well as the inevitable discussions with special branch police, government ministers, anti-insurgency theorists, and officers of various armies—forms the foundation for this work. There *are* extensive conventional sources—a brief survey can be found in the bibliography at the end of this book—but the crucial sources are largely people. The subject remaining is a delicate one, so that authoritative—much less, definitive—study must await future solutions or failures. At the moment most of the adventures of the Croatians or the South Moluccans can be found in the newspapers and in the immediate recollections of those involved. Hence for the past two years, apart from reading the newspaper accounts of terrorism, I have been engaged in talking to those involved—in the Irish countryside, in government offices, Roman basements, police stations, and on dimly lit street corners. Such sources are difficult to footnote, but they are vital to a basic book on democratic responses to terrorism.

In this case, other than the comments and observations of my revolutionary acquaintances in and out of democratic societies—who all take a keen interest in such matters—I have also contacted a wide variety of people opposed to the gunmen. Their involvements in the new terrorism have been varied: Sir Robert Mark of Scotland Yard; Judge Giuseppe di Gennario, kidnapped in Rome by the *Nuclei Armati Proletari;* Dr. Dick Mulder, who negotiated with the South Moluccan extremists in the Netherlands; and British officers stationed in Northern Ireland. I have also attended a seemingly endless series of academic conferences; conferred formally and informally with government officials, especially in Washington; presented papers; given evidence before Congress; consulted with insurance lawyers and those responsible for nuclear safeguards; and gone over the more pressing problems with those involved, from A. M. Rosenthal, then Managing Editor of *The New York Times,* to those in the United States Treasury Department who try to close down the illicit arms trade. And several people have read various manuscript drafts, including the Croatian hijackers whose adventures open the book.

A few very special, personal thanks are in order. First, thanks are due to Richard Ware of Ann Arbor, who with a Relm Foundation grant underwrote a trip to Ireland and Italy for me to begin this work and then, as President of the Earhart Foundation, arranged funding for the production of the manuscript. Clearly, he is a most perceptive foundation director. In Italy Professor Carlo Schaerf and Dottore Alessandro Silj of the International School of Disarmament and Research on Conflicts (ISODARCO) have been most cooperative—and genial—during the course of ISODARCO's investigation of Italian terrorism. And my very old friends Sigi and Francesca Maovaz, as always, make any Roman visit a joy. In Dublin Oliver Snoddy is and has always been a mine of useful information and appropriate contacts. In the United States I have benefited from or been driven to distraction by the wit and wisdom of my academic colleagues. And the congenial atmosphere of the Institute of War and Peace Studies at Columbia, the charm and efficiency of Anna Hohri, and the support and interest of Director Warner Schilling and former Director William T. R. Fox have greatly eased the problems of writing a basic book about matters on which there is no basic agreement.

J. BOWYER BELL

Rome
New York
Dublin

June 1977

PART ONE

A TIME OF TERROR

O, I have pass'd a miserable night
So full of ugly sights, of ghastly deamons
. . . .
So full of dismal terror was the time.

Richard III,
Shakespeare

Today terror, a form of political violence that falls between war and peace and offers a model to madmen and criminals, appears all but endemic in open, liberal societies. State terror, of course, has long been with us, even if only as a relatively rare option for democratic governments. The present threat, however, has been posed by revolutionaries who murder in the name of a higher law and often an arcane cause, and who use the very freedom of Western societies to stage their violent spectaculars. Here too there has been a tendency to forget the long, bloody history of Western political violence: the assassin's toll, bombs tossed into theaters, landlords murdered, factories burned, lynch mobs, and urban riots. Instead, the many and the media perceive the dramatic slaughters of recent years, the machine-gunning of innocents, the no-warning bombs, the murdered diplomats, and the extended hijacking odyssey, as novel and dread threats that must be met with novel and effective responses to defend the freedom and liberty of the West, now so inexplicably threatened.

For terror, however defined, has most assuredly now become a serious Western preoccupation.

Although killing for political purpose by rebels against the order of the day has always existed, even within open societies capable of accommodating demands for radical change, the new terror is—if only because it is perceived as such by the threatened—new. To many the spectacular massacres appear fruitless displays by men and women frustrated beyond reason, not rational political acts that can lead to power, but horror on horror's head that gains not power but publicity, that in fact is counterproductive. And this at times may be so. Most certainly everyone recognizes that a complex, postindustrial society is also an open and vulnerable society where a very few determined fanatics can cause disproportionate chaos. A handful of true believers can steal an airliner worth tens of millions of dollars, an airliner filled with hundreds of hostages, and—under the eye of the media, instantly transmitting the choreographed drama to hundreds of millions—play out a terror spectacular that may last for days and stretch out over several continents. These, transnational terrorists, the television terrorists, are indeed a new phenomenon.

Yet most revolutionary violence is neither especially novel nor inexplicable. Too often, however, those who are threatened and those responsible make no special distinctions between the madman, the criminal, the vigilante, and the rebel with a cause. For them, a bomber is a bomber, a gunman a gunman. But the psychotic bomber or the hijacker is more often than not simply mimicking the revolutionary fashions of the moment. It is the intention and potential of the revolutionary—rational, if desperate and willing to take disproportionate risks with lives (including his own)—that present the real threat. In an era of high technology, this threat might not be simply spectacular, a media event, but lethal on a grand scale. The new terrorist could not only employ the more deadly aspects of high technology to hold whole nations to ransom, but, by encouraging authoritarian repression, he could also warp the freedom of all.

Thus almost out of nowhere came fanatics dedicated to strange causes, manipulating the vulnerabilities of the West. That these new fanatics might arise from old quarrels and employ in part traditional methods did not seem as important to the threatened as did the realization that in a troubled world a new and dangerous phenomenon was putting the freedom of all in jeopardy. Suddenly, unexpectedly, the West entered a dismal time of terror.

In this time of terror no recent event so reveals the nature and scope of the threat, the vulnerabilities of Western society, or the limitations of existing safeguards as does the hijacking of TWA Flight 355 by five Croatians on September 10, 1976. Over a single weekend the governments of the United States, Canada, Iceland, Great Britain, and France had to cope with a continuing and highly dangerous terrorist spectacular, mounted by five obscure, self-proclaimed freedom fighters who penetrated airport security, dictated terms to airlines and governments, and caused the death of a police officer. They captured not only Flight 355 but also the media—newspaper headlines and prime-time broadcasts. They demonstrated that modern, highly complex technological societies are vulnerable to even the ill-trained and ill-prepared terrorist. Most of all, their adventures put the spotlight on national divisions of opinion concerning an appropriate reponse to terrorism, the real limitations on those responsible, and the present inability of those involved to reach a consensus on a proper reaction or even the meaning of the events unfolding across continents and on the television screen.

CHAPTER 1

THE CASE OF THE CROATIANS

"Tell me, please, what are we being killed for?"
Captain Richard Carey,
TWA Flight 355

At 6:30 P.M. on Friday evening, September 10, 1976, the last passengers for TWA Flight 355 from New York to Chicago's O'Hare Airport were still filtering past the LaGuardia Airport security check. The only curious thing this evening had been two gray cooking pots, weighing about three pounds each, carried on as part of hand luggage by four young men. The pots had been X-rayed but were empty, and a hand-held magnetometer turned up nothing suspicious. The four explained simply that they had bought the pots in New York and were taking them home, an explanation that seemed likely enough. The security attendants turned up nothing else of note. They rarely did.

Airport security seemed to have improved. After the hijacking of a Southern Airways jet by three criminal fugitives in November 1972—and a nine-stop odyssey through the Midwest, Canada, and ultimately Cuba—federal authorities had instituted tighter airport security measures. Beginning in January 1974, there was strict surveillance at the departure point—passengers had to walk through an arched magnetometer that signaled heavy metal objects (as well as too many coins or metal combs), an X-ray machine was used to check hand baggage, and an armed guard was stationed in the departure area. Airlines personnel were also provided with a psychological portrait of potential hijackers, a portrait sufficiently effective to weed out most psychopaths.

As for potential air pirates, once Cuba had begun to extradite hijackers under a bilateral agreement with the United States, they seemed to have given up trying to snatch airliners. In fact commercial airline hijacking had actually been eliminated by September 1976. Security checks had become routine, another of contemporary life's inconveniences. The attendants barely glanced at the X-rays, and the magnetometer's results were usually negative. The five-man Mohawk Maintenance Company team at LaGuardia expected no trouble with TWA 355, and there wasn't any. TWA had been assured that security was always stringent at LaGuardia—and more so since December 1975 when a bomb had detonated in a luggage locker, killing eleven people and injuring seventy-five others.

The last of Flight 355's eighty-six passengers boarded the three-engine 727 and strapped themselves in; at 6:45 P.M., Captain Richard Carey was cleared for takeoff. He guided his plane down the runway on what is normally one of American aviation's greatest milk runs—off at 6:45 P.M., down at O'Hare at 8:03 P.M., straight up, straight down, eighty-eight minutes, no exotic ports of call, no long hops. Except this time.

Soon after the "no smoking–fasten seatbelt" signs had been switched off, one of the passengers, Robert Goldstein, a twenty-nine-year-old management consultant from Manhattan, noticed several men moving to and from the rear lavatory. One, complaining of air sickness, locked himself in the toilet and would not come out. Sam Edsall, another passenger on his way to navy boot camp, who was sitting in the last row, observed nothing very startling about the guy except that he had a heavy dark beard and sun glasses, easy to recognize. His name was Zvonko Busic, thirty years old. Busic was an unemployed Yugoslavian-born waiter who lived at 303 West 76th Street on Manhattan's West Side. More important, under his alias "Tige," he was the leader of four other Fighters (including his wife) for Free Croatia. TWA Flight 355 was not to be a milk run. As Goldstein later put it, "It started off as an ordinary trip. It became something else."[1]

Busic and his companion came up the aisle from the rear of the 727. The two cooking pots were doctored with black tape and switching devices—instant "bombs." They strapped on packages of clay—more "bombs." Busic handed a letter to flight steward Thomas Van Dorn. It had six parts. The first assured that the flight would be something out of the ordinary: "This airplane is hijacked." The hijackers claimed to have five gelignite bombs on the plane, and another in a subway locker across from the Commodore Hotel on 42nd Street in New York where further instructions could be found. At 7:19 P.M., approximately 120 miles southwest of Montreal and 30 miles east of Buffalo, Carey radioed that TWA 355 had been hijacked, and relayed to the Federal Aviation Administration (FAA) the Croatians' instructions about the Forty-second Street bomb. He then turned the 727 northeast toward Montreal's Mirable airport and informed the passengers over the public address system that the flight had been hijacked.

Meanwhile, the other four members of Fighters for Free Croatia moved down the aisle. Mark Vlasic, like Busic in the pilot's compartment, appeared to have turned himself into a walking infernal device with trailing wires and switches. He walked the aisle, acting the crazy, willing to die at any moment. He was not a very reassuring sight. The other two hijackers, Peter Matovic and Frane Pesut, made less of an impression on the anxious and bewildered passengers. On the other hand Julienne Eden-Schultz Busic, twenty-seven, blonde and pretty, walked up and down the aisle assuring

[1] *New York Times*, 13 September 1976.

the passengers that everyone would be all right, that there was not going to be any trouble, just to be calm. Some thought that she must have been trained as a stewardess. Nearly everyone felt better.

In Washington Ambassador L. Douglas Heck, coordinator of the Cabinet Committee to Combat Terrorism, was notified of the hijack and rushed to the tenth-floor control center at the FAA building in Washington. There had to be some coordinating center to direct the responses of the various bureaucracies involved—the New York Police Department was responsible on the ground; the Federal Bureau of Investigation, anyplace in America and up to the airplane door; and after that, the FAA, until an international boundary was crossed and the State Department had to be brought in. Over the course of the hijacking, Transportation Secretary William T. Coleman and FAA chief John McLucas appeared at the center, and the National Security Council was tied in by phone. Heck brought in intelligence and logistic experts from the State Department. There were also FAA technical experts present, including medical advisors to estimate the impact of stress and exhaustion on the pilot and crew of the 727. In addition both the FAA and the State Department had psychiatrists and psychologists present to assess the probable response of the hijackers.

At 7:45 P.M., TWA 355 landed at Mirable without incident. TWA officials hurriedly made appropriate arrangements for the next stage of the hijack. Carey and the crew were not cleared for international transatlantic flight; the three-engined 727 could not make it all the way to London, the apparent destination, in one hop. Refueled at Mirable, the 727 would fly on to Gander in Newfoundland and TWA would make arrangements for a four-engine 707 to fly on to meet them at Keflavík Airport, outside Reykjavík, Iceland.

In return the Croatians promised to release a group of passengers at Gander if matters went as promised. At Gander pamphlets would be unloaded from the 727, to be dropped over Montreal, New York and Chicago. If not, there were always the internal devices. During the Mirable refueling, the hijackers told Mrs. Nicole Viau, the Montreal air traffic controller, that they were able to detonate the bomb in New York and "would do so if there were any false moves."[2] At the same time they had attempted to assure her that no harm would come to those investigating locker 5713; the one with the bomb. There were further instructions from the Croatians concerning the bomb and two letters of instruction. Busic's instructions made it easy to discover the locker ("Take the subway entrance by the Bowery Savings Bank. After passing through the token booth, there are three windows that belong to the bank. To the left of these windows are the lockers.") and suggested that no one would be hurt ("The bomb can only be activated by pressing the switch to which it is attached: but caution is suggested.").

In New York the FAA quickly relayed the radio message to the police. Their first move was to dispatch specialists from the Bomb Section of the Arson and Explosives Division from their drab headquarters at 235 East 20th Street in Manhattan. At approximately 9:45 P.M., other police were rushed to rope off the area around the lockers. They were followed by the press and television camera crews. Home viewers could follow the search, live and close up. The hijacking had become a prime media event.

Sergeant Terence McTigue, one of the first members of the bomb squad to reach the locker level in the subway station, sought out locker 5713. The squad decided to open up not only 5713 but also all the lockers surrounding it. Television cameras zoomed in on McTigue and the others prying away the locker doors. The atmosphere under the lights was tense. The squad's expressions could easily be read, for they were not wearing any protective gear. Finally, inside locker 5713, McTigue and the others found the promised device—one

[2] Ibid., 11 September 1976.

more gray metal pot with wires. Busic had created the bomb by following the instructions found in *The Anarchist's Cookbook,* a widely circulated do-it-yourself manual on radical tactics for potential revolutionaries. He had bought eight packages of a jellylike explosive, implanted them in a clock, and connected a detonator and a switch. The bomb squad found the switch on "off." After an attempt to fluoroscope the pot failed, the decision was made to cut the wire with a snapper and transfer the pot by protected truck to the police firing range at Rodman Neck in the Bronx.

In the meantime the two letters, typewritten on white bond paper, 10 inches by 8½, were turned over to the FBI, which found the first to be a list of demands and the second a 3,500-word manifesto of national self-determination and the Croatian cause. The first letter began:

> Both of these texts must appear in their entirety in tomorrow morning's editions of the following newspapers: *New York Times* (all three editions), *Los Angeles Times, Chicago Tribune, International Herald Tribune,* and *Washington Post.*
>
> At least one-third of each text must be printed on the first page of the first section, the remainder in the first section.
>
> Through a prearranged code word, we shall hear if these demands have been met by tomorrow's deadline. If they have not been met, a second timed explosive device, which is likewise in a highly busy location, shall be activated.
>
> The fate of many people hangs in the balance if any attempts whatsoever are made to circumvent our instructions.
>
> Fighters for Free Croatia[3]

The FBI relayed the hijackers' demands to the newspapers specified. The *International Herald Tribune* in Paris had already gone to press with the weekend edition, so editor Murray M. Weiss did not have to make any sort of decision. But the other newspapers needed to make decisions—whether to publish; whether the letter and the proclamation

[3] Ibid.

were "news"; indeed, whether the paper even had a policy for such an eventuality. A. M. Rosenthal, managing editor of the *New York Times,* thought that "If there was a journalistic policy it would be not to do it; but there can't be a policy to cover this kind of thing in every single case." In this case he decided to publish: "When you're confronted in the middle of the night, you don't have all the information, there are hostages, you've got people willing to take any risk—and the FBI is urging you to do it—it was appropriate to do it."[4] The *Times* printed quotations from the text of the shorter document in the late Saturday edition on page one, and the full text on page seven in later editions. The manifesto was printed in the Sunday issue. By then the story had become far more dramatic.

The pot device had been carried out in a steel mesh bag, loaded on the heavy bomb-truck, and driven north to Rodman Neck in the east Bronx. There, members of the bomb squad under the direction of Deputy Inspector Fritz O. Behr moved it into a sandbagged deactivation pit under heavy nylon mesh protective netting. Along with Behr were two officers, Brian J. Murray, age twenty-six, and Henry Dworkin, age forty, as well as Sergeant McTigue, a nationally recognized bomb expert who had been on the force for twenty years and on the bomb squad for sixteen years. He had regularly testified on such bombs, bombs that were "intended to draw attention—to emphasize a message."[5] While every such bomb is different—crude, rudely fashioned devices or far more elegantly constructed on the basis of special knowledge—what they almost always have in common is that no professional has been involved in their construction. The explosives may be volatile, the detonation system uncertain, the entire affair unstable. Thus, no matter how long a record of achievement—and no one had been fatally injured on the

[4] Ibid., 14 September 1976.
[5] Ibid., 13 September 1976.

New York bomb squad in thirty-seven years—the police experts must approach each bomb cautiously. The men were specialists in electronics, mechanical engineering, and explosives. All were volunteers. All knew the risks of dealing with unpredictable explosives but, like most, bomb specialists, felt that nothing could happen to them. There were risks, but they simply had to be confronted—risks are the essence of a police officer's job.

At 12:30 A.M. the pot-bomb was fitted with a remote-control deactivation device. The device was switched on and there was a sharp, fire cracker explosion. For fifteen minutes they tried to trigger the primitive bomb, but nothing else happened. After the last attempt the four waited for sixty seconds, the prescribed limit. McTigue and Officers Dworkin and Murray, followed by Inspector Behr, began to walk up to the sandbagged pit. All wore bulletproof vests. None wore the regulation protective gear—body armor consisting of heavy cloth reinforced by metallic alloys. The protective gear hampers movement and vision, and none of the men had worn it since the squad arrived at Grand Central. The standard police procedure dictated that such decisions were usually "a matter of common sense."

In this case at Rodman Neck, the men could—and apparently did—assume that the pot-bomb had already been deactivated when the wires were cut. McTigue, Dworkin, and Murray reached the pit and leaned over to remove the mesh and to see what had gone wrong. It was exactly 12:45 A.M., Saturday morning. The pot-bomb detonated with a roar, driving sand and shards of metal at high velocity directly into the faces of the officers. Deputy Police Commissioner Frank McLoughlin, standing beyond the pit, did not even hear the muffled explosion. Yet inside the pit there was carnage. Officer Brian Murray, a skilled electrician who had just filed papers for a transfer, was killed instantly. McTigue was critically injured. Both his hands were smashed, most of the fingers broken. Sand slashing into his face critically damaged both eyes. Inspector Behr and Officer Dworkin were also hit with the sand, but not as badly. The injured were removed to nearby Jacobi Hospital and an emergency surgical team began to work on McTigue. The specialists spent fourteen hours fighting to save his sight and repair his hands. Behr and Dworkin were listed in satisfactory condition.

Long before the surgeons had finished with McTigue, the pot-bomb had been carefully examined back at Rodman Neck. Apparently the deactivating device had fused the detonating circuit just enough to permit the explosion. The obvious technical concern, not only of the police but of the press, was with the bomb squad's procedures beginning with the first alarm. Common sense, skill, and years of experience had not been sufficient. Later, the head of the New York City Police Department's bomb section, Lieutenant Kenneth W. O'Neill, in announcing a painstaking evaluation of the explosion, appeared to support the decision to use only minimum protective gear. "They had made the evaluation that there was more to be gained by not wearing body armor," he said. Head hoods in particular, O'Neill said, "very, very much impair vision".[6] Brian Murray's widow Kathleen, mother of two small children, was not so sure. She claimed that the police department had failed to provide proper protection for her husband and, as a prelude to a damage suit, obtained a court order directing the department to turn over all its data on her husband's death. Quite obviously what was involved here was not so much "common sense" or applied experience but very much the confidence accumulated over years of success.

New York City has between 50 and 150 bomb incidents a year. The usual device is a small, simple pipe bomb, charged with smokeless powder and planted for thrills or vengeance. Yet every bomb remains different and each is constructed by an amateur with more or less ingenuity. From the very first

[6] Ibid., 14 September 1976

it was clear that people desperate and determined enough to hijack an airplane were unlikely to use a simple pipe-bomb of smokeless powder. Their bomb might, in fact, have the force of the locker bomb that had devastated part of LaGuardia Airport the previous December. Yet the subway level was crammed with spectators and cameramen—all potential victims, even at the roped-off distance. Apparently, while ripping into the other lockers, the possibility of disturbing a gravity-fuse or shaking unstable explosives was not seriously considered, or not seriously enough for anyone to wear protective gear. Nor was there a decision to detonate the bomb in place—on the assumption, apparently, that with the wires cut and the switch off, transporting it to Rodman's Neck was the safest course. At Rodman's Neck, however, after deactivation had failed it was clear that something had gone wrong. But why was there no remote-control video or mirror system to examine the bomb from a safe distance? Why was it necessary for four men to go up to look at the bomb? And why, of course, was no protective gear used—at least at first? The decisions had been left to the experts. There had been no routine, ritual procedure. The result this time was one officer dead, one maimed and critically injured, and two others seriously hurt.

Naturally enough, the immediate response did not focus on any police shortcomings but on those who had planted the bomb, those responsible for the killing and maiming—the bizarre Fighters for Free Croatia. At police headquarters, Police Commissioner Michael J. Codd, his face twisted in anger, told the press, "What we have here is the work of madmen, murderers." Just who these madmen and murderers might be had now become a subject of avid media interest. They were responsible for the first successful American hijack in years, and now, for the murder of a policeman. All this in the name of a "country" few except stamp collectors and specialists recognized, and in the name of an organization no one had heard of. The few that managed to get their hands on the 3,500-word manifesto, "Declaration Issued by the Headquarters of Croatian National Liberation Forces," were not greatly enlightened. It was polemic but coherent, a mix of historic grievance and special reasoning advocating Croatian liberation:

> National self-determination is a basic human right, universal and fundamental, recognized by all members of the United Nations, a right which may not be denied or withheld any nation regardless of its territorial size or numbers of inhabitants. . . . We present the Croatian issue as the issue of freedom, of a new form of cooperation. . . . We fight for Croatia, which will be, for all people, either a cherished presence or a beloved homeland.[7]

While the identity, even the number, of the Croatians still remained uncertain, the ideological background to the hijack was readily available. Croatia was the second largest component of Yugoslavia—6,500,000 Roman Catholics, often in uneasy association with the dominant Eastern Orthodox Serbs. During World War II members of the pro-Nazi Croatian *Ustasha* movement of Ante Pavelic killed 100,000 Serbs living in Croatia. His Serbian opponents replied in kind. After the war and Tito's rise to power, all political opposition in Croatia was crushed. In an authoritarian Yugoslavia that gradually legitimized Communist power, there was no possibility of a viable separatist movement and no real evidence of major or massive Croatian discontent. Outside Yugoslavia, however, there was Croatian separatist sentiment ranging across the ideological spectrum from unrepentant *Ustasha* movement advocates to antiauthoritarian Leftists. Much of the glue was supplied by anti-Communism, Catholic piety, and the recollection of old ethnic grievances. While these separatist groups were fragmented and impotent, authorities in Belgrade continued to take even such a minimal threat to Yugoslavian unity seriously—the country had too long a history of ethnic quarrels, blood feuds, and swift recourse to vi-

[7] Ibid., 12 September 1976.

olence. There were and had been all sorts of confrontations and divisions: between Christians and Moslems; between Serbs, Croats, and Slovenes; between pro-Germans and pro-Russians; and between the Macedonian Black Hand and the Serbian Chetniks.

While antiregime sentiment had long been strong in the Croatian diaspora, until 1971 there had been only one or two minor incidents; but in April 1971, two Croatians forced their way into the Yugoslavian Embassy in Stockholm, gagged and bound Ambassador Vladimir Rolovic, and then shot and mortally wounded him. In January 1972, a bomb placed in a Yugoslav airliner exploded while the plane was in flight, killing everyone but a stewardess, who miraculously survived a 33,000-foot fall in the wreckage. After that came a long series of apparently random incidents in Sweden, the United States, Germany, France—almost everywhere there were Croatians. For Belgrade the most serious incident was an incursion of nineteen potential guerrillas into Croatia in June 1972. In the ensuing gun battle all nineteen were reported killed and so were thirteen of the Yugoslavian security forces. The more spectacular Croatian operations, however, continued to be staged outside Yugoslavia. A complex hijacking and hostage-bargaining affair in September 1972 got the Croatian assassins out of a Stockholm prison and to Madrid, where they were jailed but eventually pardoned by Franco in February 1975. Yugoslav vice-consuls were shot in West Germany and in France. There were bombs placed in tourist offices in Melbourne, near the Yugoslav mission in New York City, and outside the consulate in Chicago.

It was extremely difficult to tell if all this scattered violence had any real base in Croatia. The brief Croatian spring of liberalization had been harshly squelched by Tito in 1972. And in 1975 government spokesmen announced that there had been two hundred arrests of subversives from thirteen different underground groups. But in Yugoslavia internal suppression of dissent was a conventional means of maintaining order, and the use of the words "subversive" and "underground" tended to be quite broad. The Yugoslav government still concentrated on the exile threat and the inadequacies of the host countries. And unlike many unaligned countries, Yugoslavia called for stern international action against terrorism. Apparently, the Yugoslavian security forces went further and inaugurated a campaign of international counterterror.

Beginning in 1968 various regime opponents were mysteriously slain in Western Europe. In 1976 Croatian exile sources claimed that members of their organizations were being murdered by the Belgrade secret police, although some observers felt the clandestine war was primarily an interexile conflict. In any case despite the secret war and the repeated, widespread violence—even despite the spectacular Stockholm hijacking in 1972—the Croatians and their cause had by 1976 made little international impact.

It was not only that "the cause" seemed hopeless and anachronistic; the Croatians lacked the resources and talents to publicize their aspirations—most of all, perhaps, they apparently lacked internal support. As an exile movement, isolated from real Croatian events and opposing a secure and, if need be, brutally effective government, its undirected campaign of violence had led nowhere. Those who had seized TWA 355 acted out of desperation and not without ingenuity to achieve for Croatia prominence if not power. And they reflected just how marginal their cause appeared and how limited their assets. The leader Zvonko Busic—"Tige"—was an unemployed Yugoslav-born waiter, long involved in militant Croatian politics, who in 1971 had been fined fifty dollars for aiming a firearm at another man. His wife, Julienne Eden-Schultz Busic, from Eugene, Oregon, was an unemployed English teacher who had been arrested in 1969 in a Zagreb restaurant for passing out leaflets calling for Croatian independence. The others were Frane Pesut, twenty-five, a machinist from Fairview, New Jersey; Peter Matovic, thirty-one, a football team trainer; and Marck Vlasic, twenty-nine, an electrician from Stamford, Connecticut. All three born in Yugoslavia. They were not a very prepossessing band of terrorists, and the Fighters for Free Croatia were not really an underground organization at all, only a group of like-minded militants. They had clustered around Busic, who was described by Reverend Slavko Soldo of his church, the Roman Catholic church of SS. Cyril and Methodius and St. Raphael at 502 West 41st Street in New York, as "an honest and good man, very strong in his beliefs, a quiet man who felt that life was not worth living without justice and freedom for Croatia."[8] His beliefs had taken him into the Croatian *Otpor* (Resistance) Party in Cleveland in 1969, after his arrival from a refugee camp. Then he helped form a tiny, even more militant splinter group, the Croatian Republican Party.

Exile politics is at best a frustrating and often degrading pursuit. There is little money, less hope, demeaning jobs, miserable living quarters, and a constant and fruitless round of meetings with the faithful few, poorly printed pamphlets and letters to uninterested editors, and always rumors and schisms. Busic's little group, meeting once or twice a week, was simply smaller and possessed fewer talents and connections.

At the least Busic and his Fighters for Free Croatia needed an audience. By the time TWA 355 landed at Gander, Newfoundland, they were assured of an audience beyond their wildest dreams—not just millions, but hundreds and hundreds of millions who could watch in part their hijacking odyssey in living color on their home screens, courtesy of the electronic media. There was no longer any doubt that the Croatians were news—big news. All the moral problems of publishing declarations and demands evaporated; even the manifestos had become news. At Gander the next installment of that news story unfolded.

[8] Ibid., 13 September 1976.

The 727 sat on the runway for several hours after landing at 1:00 A.M., fifteen minutes after the pot-bomb had exploded in the East Bronx. The Croatians were awaiting the arrival of a TWA 707, with its longer-range navigational equipment for transatlantic flights and a crew rated for international trips, to act as a pathfinder to Iceland and on to England. When it arrived, the leaflets were unloaded and rushed south, where arrangements were made to drop them over Montreal, New York, and Chicago. At the same time one of the Croatians moved up and down the aisle choosing the thirty-five passengers who would be released. There was no special system. He did try to find out if anyone was ill, but that was about all. Two of the passengers, Mr. and Mrs. Isaac Fenster of Sunnyside, Queens, suggested that a woman with a baby be allowed to go. The Croatians agreed. And an old man who did not speak English. They agreed. The Fensters went as well, and then the doors were closed. Captain Curtis was cleared for takeoff and the 727 trailed the 707 on the Newfoundland-Iceland hop.

The flight to Keflavík Airport outside Reykjavík was uneventful, but the strain was beginning to tell on the hostages. The long hours of sitting and waiting and wondering while the hijackers wandered about took their toll. The fact that the Croatians were "so polite it was ridiculous,"[9] that the woman acted almost like a stewardess walking up and down the aisle calming the anxious, saying that no one would be hurt because the demands "were quite simple and should be met by the authorities without any problem,"[10] did not make up for the anguish caused by the walking-bombs, men who had already said they did not care if they lived or died. At Keflavík the Croatians unloaded two more sets of leaflets and these were transferred to the 707. Icelandic authorities permitted the hijacked jet to take on sandwiches and soft drinks for the passengers. While the plane was on the

[9] Ibid., 12 September 1976.
[10] Ibid.

ground being refueled, United States Navy explosives experts and Icelandic policemen and firemen were at the airport but did not move near the 727. At about 9:00 A.M., New York time, both the 727 and the pathfinder 707 took off for the three-hour flight to London. Flight 355 had now been in progress for over fourteen hours; and as far as the Croatians knew, the operation had moved forward absolutely without flaw. They had heard nothing of the explosion in the East Bronx.

In the meantime American officials were beginning to piece together the background of the hijack and also some of the events on the 727. The thirty-five hostages released at Gander were flown to Chicago and interviewed by the FAA and FBI at O'Hare airport. But the picture remained far from clear. Some passengers reported that there appeared to be four hijackers, three men and a woman. The FAA spokesmen announced that there were either five or six, and the TWA spokesmen said four to six. Most passengers praised the hijackers' behavior. Jack Aldworth of Chicago again noted that "there was almost an excess of politeness." Jim Perlans, a twenty-nine-year-old wine importer from a Chicago suburb, gave the press the following description of the moment of takeover:

> Two of the men showed us what we thought were explosive devices strapped to their persons. One of them was a real sinister looking dude, a heavy black beard and dark glasses, and he was holding a device in his hands in front of him. The other was fingering a switch device and we could see wires going from the device into his clothing.[11]

George Riebe of Chicago said that the hijackers had not only the bombs wired to them but that they also had pistols stuck in their belts. Other passengers referred to a pistol or pistols and a Sten gun "type of thing." For the FBI and the FAA the news of pistols and grenades was disconcerting, since the

[11] Ibid.

chance of "that kind of stuff going through the screening device undetected is very, very remote,"[12] and the X-ray and magnetometer devices at LaGuardia were checked and found to be operative.

Attention focused on the woman, identified as Julienne Eden-Schultz Busic, wife of the "sinister-looking dude." Sources close to the investigation indicated that there was a considerable likelihood that she had worked as a stewardess for TWA several years ago. Neither the FAA nor security officials at LaGuardia believed that guns and grenades could be smuggled past the checks. Later in the day President Gerald Ford met with Transportation Secretary Coleman and FAA Administrator McLucas and spoke of a "breakdown" in security. He ordered an investigation of the boarding procedures at LaGuardia. TWA insisted that all prescribed and standard measures were in effect, and the Federal investigators believed that the weapons used by the Croatians might have been planted secretly on board the aircraft before the passengers boarded. Consequently, attention centered on Mrs. Busic, whose behavior suggested previous airline experience as a stewardess and the possibility of an inside operation. At the FAA control center in Washington, officials concentrated on the flight pattern of the two aircraft during the three-hour flight to England. With a range of 1,600 miles, the 727 was not going to have much airtime after an arrival over London.

Outside London at Heathrow Airport, British security forces had mobilized hundreds of policemen armed with machine guns and supported by armored cars. Special military units were alerted. While not routine, this was not the first terrorist-mobilization for Heathrow—the appearance of a hijacked airliner had become much too common at most European airports. Heathrow control tower switched into the 727–707 frequency when the two planes came within range.

[12] Ibid.

But there was no indication that either wanted landing clearance and there was no direct contact with the control tower. The controller overheard the 707 pilot say "Am on leaflet dropping run now." The 707 then made three very low passes over central London to the east. Leaflets fluttered down near Parliament and Westminster Cathedral and drifted onto the roof tops of the West End. After the last run the 707 curled up to a rendezvous at 2,800 feet over Daventry to the south. Heathrow control heard the 727 ask, "Did you accomplish your mission?" "You betcha we did."[13] Both then flew south across the English Channel toward France. The odyssey was not over.

French authorities who had been alerted for some while concerning the progress of the two planes now realized that there would be unpleasant decisions to make. The 707 might continue further but air time for the 727 was running out. Just at dusk on Saturday, the 707 came in low over central Paris for the last leaflet run. Strollers on crowded Champs-Élyseés saw the airliner swooping overhead, at what seemed less than one thousand feet, as the pamphlets swirled down on central Paris. The leaflets hurriedly collected in the streets by the curious announced that ". . . the world will not have peace until Croatia enjoys all the rights recognized for other peoples and other nations."

In the meantime the FAA was desperately trying to convince the French authorities that the 727 should be allowed to land. Fuel was running out and there were still sixty innocent people on board. The French officials at the Charles De Gaulle Airport in the suburb of Roissy had at first been adamant in their refusal to give landing clearance. They had their orders. The jets could not land. But FAA officials in Washington got in touch directly with Prime Minister Raymond Barre and he agreed to let the planes land. And so, shortly after 1:00 P.M., Eastern Daylight Time, the 727 landed

[13] Ibid.

and was directed to a remote corner of the airport near the Paris-Lille Motorway.

Interior Minister Michel Poniatowski rushed to DeGaulle Airport to direct operations. President Valéry Giscard d'Estaing and Poniatowski had determined to take a very hard line. For a time France had been relatively immune to the international terrorist incidents that had spread across Europe in the 1970s. There was a suspicion in countries less fortunate that a tacit agreement operated between the various Palestinian fedayeen groups—those primarily responsible for many of the more spectacular operations—and the French government. Certainly, the French embargo on arms sales to Israel and the subsequent Arab-French contacts had indicated that Paris was pursuing a less than even-handed Mideast policy, and might thus be rewarded.

Eventually, however, there were incidents by the fedayeen and the French proved no more skilled in countering the hijackers and terrorists than had other nations. Twice within six days in January 1975, terrorists attempted to rocket Israeli jets at Orly Airport outside Paris. The second time, eighteen people were wounded and ten hostages were taken. After eighteen hours the three members of the "Popular Front for the Liberation of Palestine—General Command" were flown in a French plane to Iraq. Then, on June 27, 1975, the French government had been sorely embarrassed by the escape of Illich Ramirez Sanchez, then known as Carlos and soon given the alias "The Jackal" by an intrigued press. When a Lebanese informer and three French security agents showed up at the door of a Paris apartment to question him, Carlos shot his way out, killing two agents and the informer, and making his way out of the country—only to resurface in December as director of a transnational terrorist group who seized the OPEC Ambassadors in Vienna and managed to spirit them to Algeria. Still other terrorists remained active in France—and at large. In October 1975, the Turkish Ambassador Ismail Erez was assassinated; and in May 1976, the

Bolivian Ambassador Joaquin Zento was shot and killed. More recently there was the case of the Air France jet hijacked over Greece and flown to Entebbe in Uganda—where on July 4, 1976, the Israelis, not the French, rescued the hostages.

After that the government decided on a new, tough, no-negotiation policy and established a special antiterrorist unit. This time President Giscard d'Estaing and Prime Minister Barre were agreed on a hard line. An hour after the hijacked 727 had taxied to the far end of the airport, Poniatowski gave the order to immobilize the plane by blocking the runway. The tires were shot out. Poniatowski then radioed the French position to the plane:

> Your plane cannot take off. You are considered personally responsible for the lives of the passengers and the crew. You have therefore the choice of two solutions: To be executed if you threaten the lives of the hostages or to surrender to French authorities in order to be immediately deported. These conditions are irrevocable.[14]

That was it as far as the French were concerned. Give up and be deported. Harm anyone and be executed. No concessions. No negotiations. To the captain of TWA 355, Richard Carey, this posture seemed insane—and deadly. All the Croatians claimed they wanted was proof that their messages had been published or distributed. He called into the tower, "Tell me, please, what are we being killed for"?[15] But Poniatowski insisted that "Only an attitude of firmness can end this kind of odious blackmail."

The ultimatum apparently did away with the problems and technicalities of arrest, interrogation, trial, conviction, and sentencing. For the authorities there was no time to quibble. The whole affair was in any case irregular, the hijackers not even formally in France. The Croatians in turn demanded to talk—by radio it was assumed—with either

[14] Ibid., 13 September 1976.
[15] Ibid., 14 September 1976.

President Ford or Secretary of State Henry Kissinger. Poniatowski would have none of that.

By then the American Ambassador Kenneth Rush had hurried from his residence to the control tower of the airport to confer with French officials. United States policy toward terrorists over the past four years had also been unequivocal and rigid—"We will not negotiate with terrorists." This was the Nixon-Kissinger posture that had arisen in considerable part out of the indignation at the murder of the Israeli athletes at Munich in September 1972 and then of two American diplomats in Khartoum in March 1973. At that time the Nixon administration, still dedicated to law and order, found such a position congenial; after all, if there were no concessions and no negotiations, the United States would not become involved in the endless cycle of extortion and hostage bargaining, new hijacking, and escalated demands. So they refused to negotiate, although the word negotiate was ill-defined. When the American ambassador in Tanzania assisted the parents of kidnapped students to contact those responsible across the border in the Congo, his anticipated transfer to Denmark was cancelled and Kissinger's displeasure made known. There was some public concern that the U.S. posture was too rigid and that at least some steps could be taken to secure the release of the hostages. The subsequent release of two Americans held in Ethiopia by the Eritrean Liberation Front seemed to indicate that American officials, although they had made no visible concessions, had in some part "negotiated" and been involved to a degree in securing their release.

Rush, apparently, had a wider brief: while refusing concessions, he could seek the safety of the hostages. His major problem was to save the hostages without "negotiating." He certainly soon knew that the letters had been published, because the Associated Press had sent photocopies on their Paris wire. He also knew that the FAA had urged the French to allow the 727 to land, not the President or the Secretary of

State. At 3:00 A.M. he spoke directly to the hijackers, indicating that he was the appropriate official. The hijackers wanted assurances that their proclamation had been published, that their leaflets had been dropped in New York, Chicago, and Montreal—and that apparently was all. Rush felt that his role above all was to assure the terrorists that this was the case. He persuaded the reluctant French officials to allow him to talk personally with Julienne Busic so that she could confirm the leaflet drops. The French agreed only to that. If Mrs. Busic left the plane, she would be arrested. She left the plane, along with a diabetic hostage. Inside the terminal she spoke with Ambassador Rush and was then permitted by the French to make three transatlantic telephone calls. One of these was to Reverend Slavko Soldo of Busic's church in New York, from whom she learned that the newspapers had published the Croatians' proclamation and that there were reports of leaflets. Then the French immediately arrested her.

Both Rush and Poniatowski now felt that the next step was up to the Croatians. The French ultimatum still stood. A senior American official laid down the line again.

> There is no question of negotiating, we don't negotiate with terrorists. There is no negotiating either from the French or American side. We gave them assurances that they got the publicity they wanted. Now the guys on the plane have a decision to make. We are hoping that they will surrender.[16]

In Washington the President's press secretary said that Rush's actions meant only an American willingness to talk and were not "negotiations."

Inside the 727 Busic and the others were reluctant to surrender. They still had the airplane and sixty hostages. And those hostages were becoming increasingly uneasy. Well before the landing in Paris, the strain had begun to tell on them. One of the passengers, the Right Reverend Edward O'Rourke, the Roman Catholic Bishop of Peoria, Illinois, felt

[16] Ibid., 12 September, 1976.

that his position made it encumbent upon him to play a part in the drama. He first sought to convince the hijackers that "no matter what the cause, it was a sin to encroach upon the rights of others." When that failed and the hours dragged on, Bishop O'Rourke went forward and took over the intercom from the cockpit and addressed the passengers. He gave them all absolution and told them it was time to make their peace with God. Not all the passengers were equally thankful. Gary Grecco, a Las Vegas television producer, told Bishop O'Rourke that it was not right to instill fear in the passengers. He should offer encouragement, Grecco said. The bishop insisted that because of the seriousness of the crisis, it was time for the passengers to prepare the spirit for the afterlife. But many did not want their spirit prepared and, fearing panic, flight attendant Basia Reeves tried to get the bishop to stop depressing the passengers. But the bishop proceeded anyhow and led the spiritually minded in prayers.

All of the remaining passengers were now ordered to move up to the front of the plane and around the hijackers' bombs. There, with the passengers huddled tightly around the explosives under threat of death, the bishop began the last rites. They could only wait out the ordeal. The Croatians had permitted one of the passengers, a Mr. W. Knudson, to leave the plane accompanied by a steward. He told reporters that "I am afraid for the other passengers. Help us. They are going to kill everyone on the plane."[17] It was 4:00 A.M., Paris time, and in the control tower Rush and Poniatowski were waiting out the Croatians. They were not going to go anyplace. The French had shot out the 727's tires.

At a little before 7:00 A.M., the Croatians gave it up. Busic flipped the switch on his "bomb." Horrified, the pilot ducked. There was nothing but a click. The passengers were giddy with relief. The Croatians tore apart their "bombs"

[17] Ibid.

and gave the released hostages bits of the plastic putty as souvenirs that the French security officials later seized. Given a choice of being deported to Yugoslavia or back to the United States, the Croatians chose the latter. The French immediately began preparations to fly them out to New York in a military DC-8. Since they had never officially been in France and their arrival was "irregular," formal extradition proceedings could be waived. Before they were hustled onto the DC-8, the hijackers managed to make a final statement to the reporters.

> We are proud of what we did. Don't be surprised if you hear about other attacks in the future. We are defending a just cause and yet here we are with handcuffs on our wrists.[18]

After thirty hours all that remained to be done was to tidy up the loose ends. The passengers, except for one who had complained about the bishop, were flown back to the United States. They seemed not much worse for the wear, exhausted and relieved. Like most rescued hostages, they had kind words for their captors. Warren Benson, director of the Arthritis Foundation in Tucson, was typical. "I wish them well. They had nothing against us, but wanted only to get a story across. They were concerned for our welfare, and we were treated well during most of it."[19] He felt there had been no real resentment against the hijackers.

This was hardly the case back in New York, where state and federal prosecutors were preparing indictments. Busic said that he alone was responsible. But no one was buying that. All five were responsible. Forty FBI agents along with the New York police and customs officials met the French DC-8. The police had lost one of their own and he was just as dead whether or not the bomb had detonated accidentally. The five were indicted for second degree murder by the State

[18] Ibid., 13 September 1976.
[19] Ibid.

of New York and, by the Federal government, for air piracy and causing the death of a New York City police officer.

The New York police called off the search for a second bomb. The FAA no longer had to worry about how weapons were smuggled on the plane, since there had been no guns used. The FBI and the New York police accepted that the Fighters for Free Croatia consisted of no one other than Busic and his colleagues, at best a "loose-knit organization" and at worst five fanatics with almost no resources. Assistant United States Attorney Thomas R. Pattison noted that the extent of their resources was not known. Still, quite an amount was needed for the hijacking alone. Yet no other crimes were uncovered and, as the press pointed out, five one-way fares from New York to Chicago at eighty dollars each would total four hundred dollars. The plastic putty and wires for the bomb would hardly have cost much. And the pamphlets had cost nothing, since they had been taken from the office of the International Croatian News Agency.

Spectacular terrorism simply requires a willingness to take a disproportionate risk with lives—your own and those of the hostages. As the murder complaint noted, "the defendants, acting together, acting under circumstances evincing a depraved indifference to human life, recklessly engaged in conduct which created a grave risk of death to another person and did thereby cause the death of Police Officer Brian Murray." They also, of course, had put under grave risk of death the lives of crew and passengers. And it was that drawn-out, thirty-hour ordeal that had focused and thus exaggerated the attention of the press and the broadcast media: a real-life, violent drama with an unknown ending. It was not *simply* the violence, but also the suspense. On the front page of the *New York Times* on September 11, the hijacking was the major story, spread over three columns with flash headlines. On the other side of the paper, with subdued headlines, was a far more violent, but less dramatic story: ALL 176 ABOARD DIE AS 2 PLANES COLLIDE ABOVE

YUGOSLAVIA. It was the worst midair crash in history but without the same dramatic tension as the hijack.

Some supplied the rationale for the Croatians—"With no intention to injure anyone, five sky-jackers now face death sentences as a result of their attempt to get some facts published in the press."[20] Others criticized the "cause"—". . . a group of disoriented youths who have been systematically deceived and indoctrinated by the remnants of the old Pavelic Ustashi bands. . . ."[21] And there were those pleading the middle way: "No doubt the Croatian people have been persecuted under the Tito regime . . . no one has the right, for any reason, to recklessly endanger the lives of innocent human beings, as did the Croatian skyjackers."[22] For his part, President Ford stressed the need for international action. There was concern over the French hard line and the State Department's no-negotiation policy. The *Washington Post,* however, editorially stated that it was encouraged by the French hard line and the events at DeGaulle Airport, for they revealed "that a government determined to act like a government has a good chance of finding appropriate means. Fortunately, the struggle against terrorism does not have to await the achievement of an international consensus at the United Nations." A great many Americans had become nearly as indignant with the UN's actions and lack of action as they were with the terrorists. In any case, after the hijackers' return to New York, interest in the story waned. Outrage and indignation are momentary emotions.

The most outraged party was the Yugoslavian government. For some years Belgrade had been putting pressure on countries with a large Croatian exile population to monitor their political activities. Belgrade protested to Washington that the FBI had failed to stop the terrorists and, worse, had complied with their demand that anti-Yugoslav literature be

[20] Ibid., 3 October 1976.
[21] Ibid., 27 September 1976.
[22] Ibid., 20 October 1976.

distributed. Tanjug, the Yugoslav news agency, reported that "Plane hijacking in the United States reactionary quarters, which opposed the development of friendly relations between the United States and Yugoslavia."[23] There was a feeling that if the five had been "Communists" the FBI would have followed their actions more closely. In turn the United States government, while reaffirming its opposition to terrorism, explained to Belgrade that people cannot be prosecuted for their political opinions. The Yugoslavs remained unconvinced and unhappy with Washington. At the very least in the case of the Croatians, the American authorities *had* negotiated with terrorism, *had* facilitated publication, and had thus made concessions. State Department officials insisted that this was not so—there had been no real negotiations or concessions. And anyway, once the plane was down at De Gaulle Airport, it had all been a French problem, with Ambassador Rush giving only moral support from the control tower.

As for the French, they felt that they had solved the problem rather neatly. No one had been hurt, and the guilty were dispatched for punishment. Other observers were not so sure. John Corris of TWA had different priorities than either the French or American government and made a different analysis. "We always felt that if we met their demands for leafleting and getting them to Europe, they would release the hostages. And it worked."[24] The French hard line had added to the risks. The French posture had seemed as much a result of previous failures and of concession to other terrorists as it had reflected a reasoned approach to the hijackers. No one really had any idea what the Croatians might have done—certainly there was evidence that other political hijackers had in the past responded with violence. There was no real course record; but once the 727 had been disabled, there was no going back. And at that it appeared that without Rush's

[23] Ibid., 14 September 1976.

[24] *Washington Post*, 14 September 1976.

prodding the French would not even have allowed the hijackers to verify the distribution of their manifesto. This certainly was a very hard position and, as the 727 captain had noted, an unreasonable one: "Tell me, please, what are we being killed for?" And at that point the only answer seemed to be French pride.

The French claimed that there had never been any need to worry once it became clear that the Croatians were involved and there was no possibility of a haven or revenge, as was usually the case with Arab hijackers. They also claimed that during the negotiations they discovered that the hijackers were acting alone. There could be no subsequent hijacking to free the Croatians. Just *how* the French discovered this before they delivered the ultimatum and shot out the 727's tires was not made clear—after all, the Stockholm assassins had been freed by fellow Croatians in a subsequent attack. Finally, the French claimed that the Croatians were different from the Palestinians, more malleable and logical, apparently, and unwilling to detonate their bombs. This, too, seemed to imply a great deal of French knowledge about the psychology of the Croatians—although it was certainly true that some Palestinians had detonated their bombs. Mostly, however, the French were not interested in discussion. Their strategy had worked. This time anyway.

Others had long sought different ways to make such terrorist operations if not impossible then at least more difficult. Nearly all the Western democracies favored some form of international sanctions—and so too, for that matter, did the Yugoslavs. But as the *Washington Post* observed, all efforts to reach a consensus in the United Nations had foundered. Many Third World countries, especially the Arab states, feared that the powerful simply wanted to take away the weapons of the weak and cripple their freedom fighters.

Although on the agenda of the United Nations General Assembly since 1972, "terrorism" had proven a divisive issue. There had been no action for two years, "owing to lack of time". After the Croatian extravaganza, the leading Western European governments—at the initiative of the West Germans—prepared for submission to the General Assembly a formal legal convention against the taking of hostages. The Europeans felt that whatever the wisdom of the French tactics on the ground, what was needed was an international agreement on at least this one category of "terrorism"—one that could now be proposed in the wake of an incident that had nothing to do with Third World or Arab aspirations. And this time the Yugoslav position could be a real asset in the General Assembly. While it was comforting that the Croatians had not escaped, it was most important to prevent others from employing a similar strategy.

Indeed, what was most distressing was that such a "strategy" worked. Five not especially talented people with limited funds and unlimited daring had held the attention of much of the world's news media. And the media even had a few second thoughts about their own response. Everyone agreed that there was no firm rule on such matters—publishing under threat. William F. Thomas, editor of the *Los Angeles Times,* summed up what many felt:

> We had no problem making our decision. Anybody is fair game for someone who wants to kill people. What keeps getting ignored here is that the death of the policeman, with others in critical condition, made what they were demanding a legitimate news story. There is no doubt about it. I'd react the same way.[25]

What Thomas and a good many in the press and broadcast media tended to miss was that it did not really matter what the newspapers printed or what editorial comment television stations used as a voice-over for the live film. What the Croatians wanted was exposure, measured in column inches and in broadcast minutes. And they got just that for three days. For thirty hours, five unknown people had monopolized the attention of millions upon millions and had transformed the

[25]*New York Times,* 14 September 1976.

name of an obscure Balkan province into a household "issue." For them, terror worked and worked spectacularly. The lesson was not lost in the West. After nearly a decade of increasingly spectacular incidents, more often than not choreographed for the media, there was still no consensus on the appropriate democratic response to terrorism.

CHAPTER

THE STRUCTURE OF VIOLENCE

What matter the victims, provided the gesture is beautiful?
Laurent Tailhade

There seems to be no easy answer to the threat posed by men who by waving a kitchen pot could hijack a jet filled with innocent people. There seems to be no sane way to placate those who would risk so many lives to publicize such obscure and improbable causes. And to many in democratic societies, it increasingly seems as if there is no hope that the world can long withstand the random assault of the new terrorists.

Global order, of course, has always rested on a delicate web, easily shaken for profit, out of perversity, in the name of need, or for national interest. There have always been danger zones beyond the law—pirate coasts, bandit country, the edge of the wild province filled with restless natives. Nor have supposedly stable and peaceful societies been spared the men of violence, the assassins and bombers for justice, the vigilantes or revolutionaries dedicated to a higher law. All the civilized institutions of international order have had a troubled history: ambassadors maimed, piracy encouraged, revolutions sponsored, travelers murdered—and always we have had the violent interruption of war, an excuse and explanation for all things, most of all chaos and injustice.

To fashion the rules of war and diplomatic practice, to protect the deep-sea lanes and the rights of passage, to substitute the habits of law for recourse to coercion, to encourage the reduction of separatism and parochial suspicion has long been the ambition of idealists. And by mid-century this had, increasingly, been done, if not through religious conversion or moral force or even disinterested example, then by necessity, through the play of market forces or the iron laws of technology, as a result of a balance of terror or the logic of greed. Although well over a hundred different banners fluttered over the seat of world order at the United Nations, the nations of the real world had become increasingly interdependent, visibly and invisibly woven into a complex transnational net of institutions, attitudes, and agreements. By the early 1960s, wars somehow were seen as limited aberrations. The old empires were going, often peacefully. Except for the growing quagmire in Vietnam, the new wars were short, if bloody.

Then came the assassinations, kidnappings, hijackings and massacres. As most historians who wrote on the new terrorism pointed out, the techniques and tactics were hardly new; revolutionary violence in the name of national liberation or a radical society had been about for nearly two centuries. Indeed, if there has been in modern times an archetypal terrorist organization dedicated to revolutionary violence, the People's Will (*Narodynaya Volya*) of Tsarist Russia most nearly qualified. Its members, Russian populists frustrated in their efforts to organize an indifferent peasantry, advocated a philosophy of "personal terror." After several attempts they ultimately succeeded in March 1881 in killing Tsar Alexander II by detonating one bomb under his sled and a second under him when he jumped down to investigate. The tiny People's Will, which in 1881 had perhaps as few as fifty activists, became the model for many other terrorist groups, particularly the revolutionary anarchists, who would kill and bomb for two generations. The most extreme theoretician was Serge Nechayev, protege of the revolutionary anarchist Mikhail Bakunin. Nechayev wrote a *Revolutionary Catechism,* although he founded only fantasy organizations and never became involved in a serious revolutionary deed. For many, however, the *Catechism* characterized the archtypal terrorist:

> The revolutionary is a dedicated man. He has no personal inclinations, no business affairs, no emotions, no attachments, no property, and no name. Everything in him is subordinated towards a single exclusive attachment, a single thought, and a single passion—the revolution. . . . he has torn himself away from the bonds which tie him to the social order and to the cultivated world, with all its laws, moralities, and customs. . . . The revolutionary despises public opinion . . . morality is everything which contributes to the triumph of revolution. Immoral and criminal is everything that stands in his way. . . . Night and day he must have but one thought, one aim—merciless destruction . . . and he must be ready to destroy himself and destroy with his own hands everyone who stands in his way.[1]

Were the new revolutionaries—the Croatians of TWA—such men with no names? To many it seemed absurd that these figures out of old Russian novels should have a relation to the Croatian "liberation struggles." And opponents of hijacking and hostage-taking agreed—the new men were novel and vicious, no heirs to a long revolutionary ancestry. The new terrorism was indeed new.

By 1976, a change in attitude had taken place. As late as 1969 the *New York Times Index* did not even include an entry for "terrorism"; nor had it in previous years. But by 1976 when TWA Flight 355 was diverted, everyone knew the meaning of the word and that the Croatians were terrorists, even if they claimed to be simply defending a just cause. Hijackers, kidnappers, and assassins might claim to be freedom fighters or national liberators; but by 1976 much of the

[1] Quoted in David C. Rapoport, *Assassination and Terrorism* (Toronto: Canadian Broadcasting Corporation, 1971), p. 79.

Western democratic world no longer saw such deeds as anything but descents into barbarism. There were always defenders, of course, for the Palestinians in the Arab world, for the Irish gunmen in the Celtic diaspora, for the German or Italian radicals among the chic Left. For most in democratic societies, however, terror—like love—was easy to recognize, if difficult to define.

What is terror and who are its practitioners? As all the revolutionaries and radicals and much of the liberal establishment have pointed out, quite accurately, the prime seat of terror is the state. Some regimes rest literally on institutionalized torture, vast imprisonments, random murder, and authorized gunmen. Soviet Russia had its Gulag Archipelago, François "Papa Doc" Duvalier of Haiti had his dreadful gunmen of the *Tontons Macoutes,* and the Greek colonels had their state torturers. Hitler was an arch terrorist, murdering millions upon millions; Mussolini and Franco or Mao and Stalin—these were the real terrorists. Even democratic America had killed the innocent in Hamburg and Dresden, Hiroshima and Nagasaki.

But while it is true that revolutionary terror cannot be studied separately from the far more awesome and extensive state terror, most of those examining the phenomenon know that the focus of the new terrorism is the nonstate actor—the criminal, the psychopath, the vigilante, or the revolutionary. Unlike even the state's assassination team or an intelligence service's bank robbers, new terrorists use "unauthorized" violence. And unauthorized violence, especially when spectacular, has a disproportionate impact.

The Criminal as Terrorist

There are few master criminals. Murder is most likely to come at familiar hands—and, if not, is no more probable statistically than choking to death. While there is an inclination on the part of most law enforcement agencies to consider a wide spectrum of antisocial behavior as "terrorism," the sane criminal exploiting his limited talents can be regarded as a terrorist only to the extent that he mimics revolutionary tactics and cloaks his activities with revolutionary rhetoric. Indeed, a most important indicator of social decay is the difficulty that both police and perpetrator encounter in distinguishing between criminal and political acts. When revolutionaries began kidnapping symbolic victims and demanding revolutionary ransoms in increasingly large amounts, the attention of the self-interested was instantly attracted. In Italy, for example, kidnappings occur at a rate of over one a week, bodyguard insurance and security services are avidly sought, and only rarely are politics involved. In America the realization by wanted men that by waving a water pistol they could avail themselves of a ten million dollar jet escape vehicle to transport them to sanctuary in Havana in part explained a growing number of hijackings. Criminals are not terrorists, just creatures of fashion with an open eye on the main chance.

The Psychopath as Terrorist

Although the term *abnormal behavior,* like terror, has continually defied satisfactory definition, it seems clear that those who attempt bizarre, ostensibly political actions with uncertain or irrational motivations do so for what are in fact internal, personal reasons. America in particular has been regularly the scene of a variant of deviant behavior, the psychotic assassin. A distressing number of Americans make and/or attempt to carry out threats against significant political figures, particularly the President. The assassin's profile is well known. He (or she, since the blundered attempt on

President Ford in September 1976 revealed that such an act is not sex-linked) is usually the product of a broken home, has serious sexual problems, has a history of frequent unemployment, and is incapable of establishing effective personal relationships. Such a profile, however, is all too common to be of practical use. Even a few thousand such potential psychotic assassins in the United States, a nation of more than two hundred million, would and does defy the preventive capabilities of the Secret Service. And the demented are by no means concentrated only in the United States. For example, on November 27, 1970, at Manila Airport, a Bolivian painter named Benjamin Mendoza y Amor Flores attempted to stab Pope Paul VI with a foot-long knife. His explanation, printed in the *New York Times,* was the typical babble of the demented:

> . . . Because it's time. . . . Because it's time to break down any kind of superstition and I believe my conviction was planned a long time ago in favor of people showing, perhaps, a better way of living and a better world, thinking there is a reality at two times, is something different, totally different. Crime, crimes such as Vietnam, nobody can punish, never perhaps, they will be punished. Power is helped by great superstition which is the Christian religion.

The explanation is ordinarily couched in the jargon of the day. Giuseppe Zangara, who killed Mayor Anton Cermak of Chicago while attempting to assassinate President-elect Franklin Roosevelt, claimed that the hatred of kings and presidents was his motivation. Sirhan Sirhan said he killed Senator Robert Kennedy out of love for Palestine. In the nineteenth century, an attempt on the life of Andrew Jackson was made by a destitute house painter who claimed to be Richard III of England.

If the assassin has always been with us, obviously the aerial hijacker has not; but the opportunity to be involved in such a spectacular deed soon attracted the disturbed. These were largely men with certain highly recognizable and remarkably similar personality traits and personal backgrounds. Some, of course, were criminals, but most were simply disturbed, attracted by the power of command and the prospect of capture after a life of frustration and failure. Both groups tried, haltingly, to patina their act with revolutionary rhetoric in order to hide their fantasies behind a curtain of fashionable reality. Some hardly bothered. On February 22, 1974, an unemployed salesman who had previously been committed to a Philadelphia hospital for mental observation and had twice been arrested for picketing the White House, killed a guard and a pilot in a bungled attempt to seize an airliner in the Baltimore airport and crash it into the White House. Later that year, the White House was similarly threatened by a helicopter assault and then by a fake "human bomb." And the bizarre hijacker need not be an American. In August 1973, for example, Mohammed Touni, a Libyan citizen with a history of mental problems, took over a Lebanese jet and ordered the pilot to fly to Tel Aviv "to bring peace" to the Middle East.

In many cases the mode of violence is selected from the fashions of the day—assassination, hijacking, hostage-taking. In January 1977, in Los Angeles, twenty-one-year-old Dolpin Lain seized and held a hostage on top of the city's tallest building, the sixty-two-story United California Bank Building, in order to gain publicity for his nonsmoking campaign—his father had died of cancer. And there was obviously *some* method in his madness—the *Los Angeles Times* printed a long interview entitled "Stop Smoking: The Media Got His Message." In suburban Cleveland ex-marine Cory C. Moore seized police captain Leo M. Keglovic and among his demands insisted he speak to the President and that all white persons vacate the planet. (Forty-six hours of negotiation led to the release of the hostage and a telephone call from President Jimmy Carter as part of the bargain.)

There is a tendency on the part of some, including law enforcement agencies, to view any such act of violence as rational, criminal behavior. In general Americans seek rational explanations, the comfort of reason. Thus the young Dolpin Lain was driven to act as he did because his father had died of cancer, and President Carter felt he must keep a "bargain." It has been comforting for the many to explain President John Kennedy's death as the result of a political conspiracy. There has consequently been a blurring in the public mind between rational "terror" and aberrant behavior, and in the mind of many law enforcement officials, between criminal terror and psychotic acts.

Violence as Therapy

Manson and the Hanafi

Only when the behavior, motivations, and rationalizations of those involved are so obviously aberrant that no rationalization is possible has there been a refusal to take the "terrorists" at their own valuation. The strange Charles Manson cult in California, whose members explained their murders as an effort to begin a race war—Helter Skelter—attracted vast interest because their deeds were so violently pointless. They had killed without reason, without remorse, without effect. They terrified without being terrorists.

On the other hand, in Washington, D.C., in March 1977, when the small Hanafi, Black Moslem sect seized 135 hostages, there was a tendency to regard the act as deplorable and criminal, but rational. Hamaas Abdul Khaalis' family had been slaughtered in an internecine, religious war. In his frustration and anguish, in his desire for absolute vengeance, in his anger that the world apparently had ignored the massacre and that the state had not undertaken appropriate punishment of the criminals, he and his friends struck out. The seizures appeared well organized, carefully synchronized, the product of rational planning by desperate means. And although no longer popular, religious fanaticism of the Hanafi was easier to understand than the Manson cult of Helter Skelter.

Yet the Hanafis were driven and fragile men whose demands were less relevant than the need for an act of cathartic violence that would ease their frustrations through the act of negotiation. Unlike the criminal, whose self-interest dominates priorities, or the revolutionary, who seeks political advantage, the Hanafis sought therapy. Here a trained police response, excellent advice, plus the carefully orchestrated intervention of three Islamic ambassadors fostered a peaceful solution. The Hanafi incident thus had little or nothing to do with religion or politics, but a great deal to do with personal anguish.

In similar cases two small groups in the United States have engaged in cathartic violence—destructive, deadly, self-destructive—two transient communities of psychopaths "explaining" their acts in uncertain, if fashionable, political terms. They have been elevated by the media and the police as America's own terrorists: the Black Liberation Army and the Symbionese Liberation Army.

The Black Liberation Army

In New York City in 1973, two police patrolmen, one white and one black, were gunned down in the East Village in a particularly brutal and apparently senseless act of murder. Two days later the United Press International received a Black Liberation Army (BLA) letter of explanation that ended with the warning that there was more to come.

This new group had evolved out of the splintered remains of the Blank Panthers. They were a collection of tiny, fluid groups of militants, continually merging and dissolving. Often recruited to black nationalism in prison, they had drifted on the edge of events while living on the proceeds of theft and robbery. Partially educated or self-taught, they had become outlaws even from a black ghetto. They were violent outcasts of the urban jungle and the prisons, driven by hatred of the system and outraged at their own limited prospects. Having learned revolutionary rhetoric, they claimed that the system had destroyed them, that the prisons, not the prisoners, were at fault.

Their past was miserable, even loathsome. They had no future. The present was a moment of desperation. For these murderous, frustrated blacks, the BLA was an acceptable self-rationale for violent revenge against authority. And for these driven men, the most visible pillar of the system was the police. They saw patrolmen, whether black or white, only as blue targets—the enemy. And so they killed, made one or two elegant revolutionary converts, killed again, and in turn were killed or captured.

The Symbionese Liberation Army

An even more bizarre "revolutionary movement" surfaced in California in 1973, when something called the Symbionese Liberation Army (SLA) claimed responsibility for the murder of Oakland Superintendent of Schools Marcus Foster. After two suspects were arrested and subsequently convicted on first-degree murder charges, the nature of the movement began to become somewhat clearer to authorities. The SLA was part cult and part ultraradical conspiracy comprised of a dozen men and women coming from diverse backgrounds—university rebels, escaped convicts, and confused drifters, lesbians, and criminals. Their black leader, "Cinque," a self-styled general field marshal, *neé* Donald DeFreeze, a thirty-year-old escaped convict, stated in a taped message that "we are savage killers and madmen . . . willing to give our lives to free the people at any cost." The SLA included killers and madmen, radicals who had stepped beyond the wilder fringe of protest and, on the fringe, a group of frantic blacks identical to those of the Black Liberation Army. The people they sought to free could be found within themselves. They were angry and frustrated. Loathing their life or feeling profoundly guilty for imagined wrongs or a previously pampered life, they had sought relief and ultimately an end to their suffering by fashioning their own "liberation front."

During 1973 their simple conspiracies and single brutal crime were of interest largely to the police. Then in February 1974, the SLA kidnapped Patricia Hearst, granddaughter of the late William Randolph Hearst—attractive, blonde, normal, a pleasing symbol of success. The SLA-Hearst episode became the media event of 1974 and much of the next year. The first SLA communique after the kidnapping was a patchwork letter filled with revolutionary jargon which warned that "should any attempt be made by authorities to rescue the prisoner or to arrest or harm any SLA element, the prisoner is to be executed." At first the Hearst family, with the cooperation of the FBI, attempted to placate the SLA. An attempt was made to accommodate the SLA demand for free food for all people receiving Social Security or participating in the food stamps program, a demand as originally delivered which would have cost an estimated 239,000,000 dollars. The Hearsts eventually put up two million dollars, agreeing to fund additional distribution if Patricia were released.

Then on April 3, to the amazement of all, Patricia Hearst sent a tape to her parents announcing her "conversion" to the SLA as Tania, girl-guerrilla. Media interest intensified. It was a live drama with no foreseeable final act. The Hearsts' repeated pleas over television increasingly took the tone of psychiatric sessions, plea bargaining with the demented. The police and FBI intensified their search.

The final act in the SLA "revolution" was a spectacular shoot-out in Los Angeles between most of the SLA members and the Special Weapons and Tactics (SWAT) squad of the Los Angeles police. Much of America practically closed down in order to watch the SLA go up in flames under the machine guns of the police. Over 7,000 rounds were expended. But Patty-Tania was not inside the flaming building. She and two other members were still at large and were not to be captured till later. Overall, the SLA war against the "Capitalist Class" had consisted of the murder of a black school official, the kidnapping of a nineteen-year-old girl, a bungled bank robbery, and a tragic-comic shoplifting attempt. The SLA members were not revolutionaries, only violent eccentrics, driven men and women living on the margin of rationality, functioning only by resorting to risk and violence that was certain to be self-destructive. They were, however, very good copy.

Such violent and irrational behavior follows fashions and cycles. General Field Marshal Cinque, the poor lost Donald DeFreeze, simply copied the rhetoric and tactics of the moment and became briefly the California Che Guevara. If Squeaky Fromme, the strange, failed assassin of President Gerald Ford, could misfire her way onto the cover of *Time*, it is not surprising that someone else—an equally confused woman with equally dim motives—would try again. Such mad deeds have no real political content even though they may, if successful, have political repercussions. The fantasies of the BLA and the SLA or the presidential assassins can most easily be explained by psychologists. Their deeds reveal only incidentally anything about the nature of society—they are merely the mimicked fashions of the moment, wrapped around dementia.

The Terrorist as Vigilante

If those who murder and maim to protect a threatened system are more unsavory, more distressing to decent opinion than the criminal or the madman, at least they are part of a recognized political process. At times the line between the two is a thin one. Often a threatened regime may prefer that volunteers rush into the breach, without formal authorization, doing for the state what the state prefers to avoid doing itself. In Brazil off-duty policemen murder "known" criminals and suspected revolutionaries—not that the state has been especially reluctant to protect itself openly by recourse to brutality. In Guatemala organizations like *Ojo por Ojo* (Eye for an Eye) and *Mano Blanca* (White Hand) regularly murder opponents of the regime with impunity. Mehdi Ben Barka and other Algerians were murdered by Frenchmen who felt frustrated by legal restraints. In Argentina it is difficult to determine whether the kidnapping and murder both of indigenous revolutionary suspects and of exile radicals is authorized or tolerated or encouraged or, in fact, official. Certain vigilantes may operate to protect a society without the support of the recognized regime. Thus in America the Ku Klux Klan sought through violence and intimidation to avoid a radical reconstruction of Southern society. At first they opposed the order imposed by the occupying Union army and filled a vacuum if anarchy lurked. Later their actions decayed into an illicit campaign of coercion to maintain white supremacy. Random murder and arbitrary intimidation of blacks became and, until very recently, remained a conventional part of rural Southern life.

In Northern Ireland after 1968, the system of Protestant predominance decayed, ending in 1972 with the suspension of the regional assembly outside Belfast at Stormont, for fifty years the seat of what the Catholics saw as institutionalized injustice. There then followed all but spontaneously a Protestant campaign of random, sectarian murder. Those Protestants who seemingly benefited least by the existing system—the urban working class and the poor—were first to defend by recourse to arms "their way of life," not simply a Protestant state for a Protestant people, but a domination real and symbolic over their "Papish" neighbors. A dozen ill-organized groups—among them, the Ulster Defense Association, the Red Hand Commandos, and the Ulster Volunteer Force—appeared, merged, split, and disappeared; but the murder of Catholics, usually men wandering on the edge of Protestant areas but rarely involved in any way in politics, continued.

When the membership of the new Northern Ireland Assembly that replaced the old Stormont seemed determined on Catholic-Protestant accommodation, the response of the desperate Protestant working class was a general strike and resort to violence. The new assembly collapsed; Britain continued to attempt to maintain order with its army; and the Protestant paramilitaries—abhorred by London, detested by their betters, representing themselves alone, brutal, crude, prone to violent schism and the lure of extortion—persisted. There was round after round of tit-for-tat murders in north Belfast, without warning, and in the countryside there was the knock on the isolated farmhouse door and the single shot. Such vigilante terror produced endemic violence in large parts of Northern Ireland. At the very least it indicated that the Protestant gunmen would hold veto power over any "democratic" system that the decent, accommodating people in London or Belfast might attempt to impose.

The Terrorist as Revolutionary

Almost without exception the public's perception of the terrorist is of someone associated with a revolutionary organization. There may be state terror or criminals and madmen, vigilantes, and authorized assassins; but for the many the *real* terrorist belongs to a revolutionary organization—to Black September or the Japanese Red Army, to the IRA or the latest Latin American National Liberation Front. For some *any* violent act by such revolutionaries, even a defensive one, is terror. Thus, uniformed guerrillas fighting a rural irregular war that has given them control over a liberated-zone and the opportunity to establish a government are still "terrorists." Others, advocates of a cherished cause, can always find a rationalization for any deed, no matter how violent and unsavory. For them *any* violent act is legitimate.

What is clear is that violence is used by revolutionaries in a variety of ways for differing purposes that may overlap but can be examined separately.

The Categories of Revolutionary Violence

Every revolutionary organization must cope with the problems of maintaining internal discipline, inhibiting penetration, and punishing errant members. Such activities take part within the organization. To be effective, punishment must be swift, harsh, and visible. It is, therefore, usually highly formalized, with a trial, defense, sentence, and execution. (It may be a sham, but it is a ritualized sham.)

Clearly, in the underground, on the run, the niceties of law and civilized restraint often suffer, if not by choice then by necessity. But generally, internal, organization terror is not meant to intimidate others, only to punish the guilty. The assumption is that most of the revolutionary apples in the barrel are sound and that the cadres can trust each other, even though vigilance must be maintained. And the occasional execution or torture session—as long as it is occasional—should have a salubrious effect on discipline. Yet in those organizations filled with fanatics on the fringe of reality, such "trials and punishments" may have little to do with organizational discipline—a girl in the Japanese Red Army was tried, convicted, and murdered for wearing earrings.

In many cases the first external evidence of revolutionary terror that goes beyond organizational violence is the revolutionaries' effort to create mass support. Such "support" may be obtained through extortion or by threats of vengeance. Sometimes, as in the case of the Algerian FLN in France, the major organizational campaign seems to rest firmly on a foundation of assassination: an Algerian in France contributed to the FLN war fund and to no other or else risked being dredged out of the Seine. In Algeria those who continued to purchase European cigarettes risked having their noses slashed off.

Since many revolutionary organizations at times resort to such acts to maintain momentum and appease militants—as well as to pay the bills and orchestrate their campaign—there has been a tendency to assume that this is the *only* means open to such movements and that whatever support exists is coerced, not volunteered. The rebels respond that without state coercion few citizens would pay taxes, serve on a jury, or risk their lives in the military. The state can imprison: the revolutionary, claiming to be the legitimate representative of the people, does not have that option.

To maintain the organization and garner support are primarily internal matters; but to seize power the revolutionary must carry the struggle to the enemy. And the enemy is defined by function: soldiers or police or intelligence agents. In Ireland, on November 21, 1920, IRA squads descended on the safe-houses of sixteen suspected British undercover agents in Dublin, killing them all and at a blow largely destroying a British Intelligence net. More often than not, the functional victim is a member of a category; and in a tightly organized insurrection in which the rebel maintains an interest in his image, there is a reluctance to broaden such target groups beyond the military or other uniformed services.

There are times when the category of victims becomes vastly expanded. *Any* traveler who flies to an enemy country is fair game. In such circumstances terror is almost random. Clearly, the major *function* of spraying machine gun fire into a transit lounge is not the elimination of the immediate victim but rather the general intimidation of all potential visitors. In fact it is rare when even the most discrete use of functional terror does not have a broader impact. The death of just one policeman may intimidate all his fellows, reveal the shakiness of a regime, and draw international attention. But for the distant observer "terror campaigns" are most easily recognized when the number and categories of victims escalate, as the targets become more random and more distant from repressive functions.

Revolutionaries, of course, can be provocative. For example, unintentionally, but not unexpectedly, when the IRA shot the British agents—a specific act of functional terror—it provoked a response that led to a long-term readjustment of the political situation and this readjustment proved to be of much greater advantage to the IRA than the deaths of sixteen suspected British agents. As part of the British reprisal, three IRA men, including the officer in command of Dublin, were picked up that same evening and shot "while trying to escape." This authorized murder of three IRA men might have passed muster in Britain—the lads had been sorely provoked by an IRA that murdered from the ditch in the dark. During the afternoon, however, a group of Blacks and Tans, an auxiliary military security force of dubious repute, had "spontaneously" fired into a crowd at a football game, killing twelve spectators and wounding sixty—Bloody Sunday. That act was one that British public opinion could not accept. Ultimately, there arose a consensus in England that if Ireland could only be ruled through a policy of counterterror, in opposition to British traditions and wont, then it was not worth ruling at all. What worked for the IRA, and later in Palestine for the Irgun Zvai Leumi, would hardly have been effective against a government less vulnerable to shifts in public opinion, less decent in nature—for example a government like that of Nazi Germany or Communist Russia.

Such manipulation is essentially strategic; it seeks to provoke a basic change in the objective revolutionary situation. Far more visible, however, has been the development of tactical terrorist manipulation. The most common use of this technique is the creation of a bargaining situation in which the terrorists threaten to destroy seized assets or hostages—

TWA Flight 355, for example—unless they are granted certain demands. The manipulation rests on the dread, if not the certainty, that such threats will be carried out. The anguish can be intensified by the visible hysteria of the terrorists, by the confusion of the moment, and by the difficulty of communication. Thus manipulative terror extends the duration of the drama of the deed while the revolutionary seeks a functional gain, in terms of freed prisoners or ransom, and *forces* the target to react in what are assumed to be advantageous ways.

Even when the target refused to "negotiate" or make any concessions, the revolutionary still wins on the manipulative level. There has been a response, and not always a popular one, by the challenged government. Not everyone at De Gaulle Airport was happy with the French response, certainly not the crew and passengers. And the American Foreign Service has been less than cheered by the realization that their leaders will not negotiate their release in a hostage situation. Even if there are no concessions, even if the hijackers end in prison, there still remains the possibility that their friends will extend the drama with another act, another hijacking and another challenge. There may, too, be spinoff benefits elsewhere if the public becomes delighted by the daring and audacity of the terrorists. Even if one target is appalled by the deed, another may respond sympathetically. Even if everyone is horrified, the terrorist drama at the least may force people to ask why men would be driven to such actions.

A final variant of revolutionary violence, symbolic terrorism, goes beyond the functional. A victim must be selected who symbolizes the enemy and then a deed must be fashioned that is so awesome and cathartic that even if history is not transformed at least the perceptions of the audience will be. Examples of such symbolic terror abound. When the *Organization de l'Armée Secrète* (OAS) tried and repeatedly failed to assassinate Charles De Gaulle, they sought not simple vengeance—the death of a man who had betrayed a French Algeria—but a symbolic triumph over France itself. When Leon Trotsky was struck down with an ice ax in his Mexican villa, not only was a man killed but a heretical movement was also maimed. When the Puerto Ricans attempted to shoot President Harry Truman, their concern was not with the man from Missouri but with the President of the United States, symbol of a nation they felt had wronged their island.

To a real degree the rising impact on the public of revolutionary violence is not a result of the number of deaths but rather of the drama of the moment, the crafted horror of an assassination or an airport massacre. This the new transnational terrorists have grasped, and this is what has made their acts novel. The revolutionary guerrilla who kills his own or who seeks popular support through the barrel of a gun attracts far less attention, especially when he does so in an obscure outback. Even in advanced Western societies low-level political murder may gradually be drained of drama—lynching may become commonplace, or the bodies at the end of an Irish lane may no longer seem newsworthy. In fact it matters a great deal not only why and how a rebel kills but also where and when. The television terrorist understands prime time, the need to escalate his deed, to manipulate the media, to reach the masses. In this, too, he is truly novel, for most revolutionary struggles have been played out on small, isolated stages.

The Arenas of Revolutionary Violence

With certain notable exceptions, the revolutionary has used his violent tactics in the course of *national insurrections,* campaigns intended to liberate an oppressed or denied people. Most revolutionary campaigns are parochial, no matter how international the rhetoric. Most revolutionary organizations are counterstates seeking recognition, insisting on legitimacy, reluctant to take up the tools of terror except when impotent to do otherwise. While from time to time there have been revolutionary alignments and alliances, guerrilla summits, and the covert exchange of views, most rebels practice their trade in isolation, suspicious of alien volunteers and suspicious of advice but comforted by distant friends.

Thus for most democratic societies the small wars of the bush, the campaigns of murder from a ditch, have been, and have seemed, far away and of little import until the sins of imperialist intervention became costly in domestic turmoil. Still, even the most obscure campaign has a tendency to ooze out and attract more general notice. No rebel is so isolated that he does not attract some sort of support and often some very strange advocates. Colonel Muammar el-Gaddafi of Libya supports the aspirations of the Provisional IRA, and the Shah of Iran, while oppressing his own Kurds, has aided those of Iraq. And the crisscrossed underground world of arms shipments, training camps, borrowed passports, dues collected in the diaspora—all spread revolutionary violence.

Even within the violence-area, the arena of revolutionary combat, the rebel may broaden the struggle without leaving home: for just as he has distant advocates, he also perceives distant and not-so-distant foes—transnational corporations, governments allied to the oppressor, foreign advisors to his opponents in the field. Thus in Uruguay British Ambassador Sir Geoffrey Jackson, a British imperialist symbol, was kidnapped, and Daniel A. Mitrione, a United States Public Safety Advisor, was shot to death. Transnational executives have become targets, murdered in Teheran, kidnapped in Buenos Aires and Caracas. In an interdependent world the revolutionary has increasingly become involved in *international terrorism*—Cuban exiles detonating a bomb in the car of the former foreign minister of Chile as he drives through Washington, D.C.; a Bolivian exile general involved in the hunt for Che Guevara killed in Paris; letter bombs dispatched to Israeli diplomats and Jewish businessmen in London. Thus, with the spread of international terrorism, violence has been injected, often for no more than tactical reasons, into open societies. Two Turkish ambassadors were assassinated in Vienna and in Paris, perhaps by Armenians, but the locations were solely for reasons of convenience. There was no message for the Austrians or French, no choice of a stage to heighten an impact. The murders really had nothing to do with Austria or with France.

International terrorism is essentially an escalation of a nation-bound struggle waged for either tactical or ideological reasons. The IRA places bombs in Manchester or London because they will have a greater impact there than in Belfast or Derry. The Palestinian fedayeen seize American diplomats in Khartoum because the United States is Israel's patron. The Croatians hijacked a TWA jet because it was convenient and vulnerable, and they ended up in France because of the limits of the plane's fuel supply.

The epitome of the new terrorism, however, is *transnational terrorism.* Here the entire complex, interdependent postindustrial world is one target. Thus the system becomes the prime enemy, not just local tyrants, symbolic ambassadors, or errant politicians. Unlike the old revolutionary anarchists who also attacked the system by recourse to personal terror, the new transnational terrorists have not limited themselves to kings and presidents or even to ambassadors and military advisors. Rather, they strike at any vulnerable node in the system: any tourist on the wrong airline, any shopper in the wrong supermarket, any citizen of the guilty states.

Since the entire system is now threatened by these new revolutionaries, there is today more general perception that the terrorism they represent is novel—and dangerous. Tomorrow will not be like yesterday. Even open, democratic societies no longer live in a surprise-free world but in a time of terror. A new and frightening dimension had been added to an already troubled world. For the most part the public and those in policy positions are not concerned with neat academic categories or analytical distinctions. They do not care how or why terror has worked—the physiology of a new political violence—only that anarchy now seems to pose an unexpected challenge to world stability. And this challenge arose swiftly as a result of a convergence of factors at the end of the 1960s.

CHAPTER

THE EMERGING THREAT, 1968–1971

I will do such things—
What they are yet I know not—
but they shall be
The terror of the earth.
King Lear,
Shakespeare

For most of the 1960s, "terrorism" had remained an historical term variously associated with the French Terror or Revolutionary Anarchism or Mad Bombers. While antiinsurgency and guerrilla warfare were fashionable policy and academic subjects, this was not true of terrorism. From time to time, of course, Western governments would label their irregular opponents terrorists. The British had fought CTs—Communist terrorists—in Malaya and the "mad" terrorists of the Mau Mau in Kenya. The French had fought the FLN terrorists in Algeria and, in the process, had fashioned a philosophy of counterterror that proved in practice elegant, rational, and self-defeating. There were terrorists to be found in Indochina and Africa—anyplace where irregular war was practiced. Such "terrorists," however, tended to wage their wars in the traditional manner of rural guerrillas, sometimes coupled with urban operations. The generation after World War II had been filled with such campaigns, beginning in Palestine in 1944 and continuing through the late 1960s in Vietnam. During those years almost every technique or tactic of the classical terrorist could be found being practiced in abundance: assassination, kidnapping, hijacking, hostage-taking, torture, extortion, massacres, and spectacular escapes or ambushes. Such violence, however, was perceived by the world in the context of an imperial war or a rural guerrilla campaign or the institutionalized injustice and repression of a threatened despotism.

There was, however, one truly novel threat to a growing vulnerability—the hijacking of commercial airliners under various pretexts and for differing purposes. The first recorded hijacking of a commercial airplane occurred in Peru in 1930, when rebels commandeered an F-7 piloted by the American Byron D. Richards and dropped pamphlets. Everyone soon forgot the incident. Immediately after World War II, the only visible reaction in Washington and the West to hijacking attempts was complacency or, in many cases, enthusiasm, for the hijackers tended to be Eastern Europeans fleeing Communist tyranny to gain Western freedom, and, after Castro's triumph in Cuba, opponents of his regime who hijacked airliners to flee the new Cuban communism. Considerable American sympathy for the underdog Cubans began to dissipate on May 1, 1961, when a National Airlines Convair 440 on the way to Key West was hijacked by a psychopath who waved a knife and diverted the plane to Havana.

Clearly, the United States was vulnerable, and so were a number of other nations and airlines. For the most part, however, the hijackers who wanted to go to and from Cuba, protest American policy on Algeria or simply fly to Arkansas to see an estranged wife proved a short-lived phenomenon. There were not really a great many hijackings, and few of those involved, innocent or guilty, came to grief. The trend toward hijackings was disturbing, especially to the airlines, but not of prime importance and apparently of minimal political consideration.

There was one special hijacking, however, that foreshadowed many of those elements that would rivet world attention a few years later. On August 23, 1966, a group of eighteen Argentine nationalists led by the beautiful blonde Maria Christina Varrier took over an Argentine Airlines DC-4 between Buenos Aires and Rio Gallegos and forced the pilot to fly to the Falkland Islands, the tiny British colony in the south Atlantic. The pilot managed to land the DC-4 on a flat strip occasionally used by the islanders as a race track. A crowd from Stanley, the capital (with a population of 1,250), gathered about the plane. Maria and the others announced that they were Condor, a group dedicated to the unity of the Falklands with Argentina. Speeches were made in Spanish to the English-speaking islanders. Flags were planted. And the news went out over the ticker, a good story that brought the existence of the Falklands and the Argentinian claim briefly to the attention of a bemused world. Then the Condor group and the pilot and passengers had to wait until the monthly mail boat showed up. Everyone eventually got home safely, if a bit late, and the Falkland issue disappeared, along with the Condor group. Quite obviously, what Condor lacked (and the Croatians had) was an understanding of the media, of the importance of the arena for violence, of the possibilities of manipulative spectaculars. Condor had choreographed a splendid event, but one without an appropriate audience. Their deed simply became a statistic, rather than a drama.

The next year there were only six hijacking cases worldwide, all of them run-of-the-mill. In 1968, however, the total increased to thirty-five and became a significant concern in the United States and to the airlines, to the government, and to the media and the general public. In America the hijackers seemed a curious mix of homesick Cubans, criminals on the run, and misfits in the grip of strange fantasies. They took control with guns or knives, with vials of colored water, toy pistols, mosquito spray, bottles of shaving lotion, and fake dynamite sticks. The Cuban connection had become commonplace, a joke to all but those involved.

But in July 1968, there was a more ominous hijacking when three members of the PFLP seized an El Al 707 out of Rome on the way to Athens and Tel Aviv and diverted it to Algiers. The PFLP people were neither criminals nor crazies, but dedicated Palestinian revolutionaries whose exploits against Israel were being overshadowed by the far larger fedayeen group Al Fatah. The crew and some of the passengers were held while the PFLP put pressure on Israel to release guerrilla captives. The spread of hijacking to an area already involved in an irregular war waged by the Palestinians, the new heroes of an Arab world shamed by the disaster of the June 1967 war, comforted none of the advocates of order. But most observers continued to see the hijacking phenomenon in Cuban-American terms.

Then in 1969 those less sanguine had their fears confirmed. At the end of 1968, on December 26, a PFLP commando team had attacked an El Al plane on the ground at Athens. On February 18, 1969, another team machine-gunned an El Al plane during takeoff from Zurich. Nor were the Palestinians the only ones. On March 11, the Eritrean Liberation Front (ELF) planted a bomb on an Ethiopian airliner on the ground at Frankfurt and machine-gunned an Ethiopian Boeing 720 jet on the ground at Karachi; on August 12, they hijacked a DC-3 on a flight between Addis Ababa and Djibouti and diverted the plane to Aden.

Still no one knew very much about Eritrea or the ELF or, for that matter, about Ethiopia. It was not until later in August that the Palestinians were able to attract full media coverage and take hijacking out of a Cuban context. On August 29, two young members of the PFLP, Leila Khaled and Salim Issawi, took over a TWA 707 flight out of Rome that was scheduled for Athens and Tel Aviv. On instructions from the terrorists, Captain Dean Carter overflew Athens and flew toward Damascus. When in range of the Lod air controller, Leila Khaled broadcast to Israel:

> We have kidnapped this American plane because Israel is a colony of America and the Americans are giving the Israelis Phantom planes . . . Tel Aviv. We are from the Popular Front for the Liberation of Palestine. What can you do about it?[1]

Tel Aviv could do nothing, Washington could do nothing, and certainly TWA could do nothing. Flight 840, renamed Popular Front Number One, landed at Damascus escorted by Syrian jet fighters. After debarkation, the plane was bombed. Most of the passengers and crew were permitted to leave Syria, but two Israeli men were held for three months and were only freed when Israel released thirteen Syrian prisoners. It was a PFLP triumph: America and Israel humiliated and impotent, and the Syrians free. A new and effective mode of striking at the imperialists had been demonstrated. There was no retribution. Israel and the United States did nothing. The International Federation of Airline Pilots Association dispatched a cable of protest to Syria but decided against a strike. The revolutionary weak had bested the "mighty" imperialists. They had also revealed a vulnerable node in transnational order—twenty-five million dollars worth of airplane and a hundred hostages could be had for the taking by the bold.

Almost simultaneously with the advent of political hijacking, there emerged two new variants of assassination that boded equally ill for a tranquil world. The murder of the mighty, of course, already had a long history—and in the United States a traumatic one. After 1945 the list of those who escaped assassination attempts reads like a Who Was Who in World Politics: Harry Truman, Fidel Castro, the Shah of Iran and Charles De Gaulle, Juan Peron, Konrad Adenauer,

[1] James A. Arey, *The Sky Pirates* (New York: Charles Scribner's Sons, 1972), p. 78.

Gamal Abdel Nasser and Jawaharlal Nehru, Mohammed Ben Bella and Kwame Nkrumah, Mao Tse-tung and Achmed Sukarno. And the peacemakers were equally cursed—the nonviolent Mahatma Gandhi and his American disciple Martin Luther King; Count Folke Bernadotte, the United Nations Mediator in Palestine; all killed along with villains like Trujillo and the advocate of apartheid, Verwoerd. Perhaps the record belongs to De Gaulle, who survived at least thirty serious plots, including a long burst of submachine gun fire sprayed into his limousine.

It was not, however, until a series of American assassinations and assassination attempts—the two Kennedys, Martin Luther King, the contentious George Wallace, Malcolm X, the neo-Nazi George Lincoln Rockwell—that political murder began to attract widespread attention. The public could not understand killers who would rather be wanted for murder than not wanted at all. It became the common wisdom of the specialists that the American assassins were in the grip of psychopathic fantasies, cloaking their deeds in the fashionable rhetoric of the moment.[2] Despite the fact that between 1918 and 1968 some sixty-eight heads of state had been assassinated, the era was not generally seen as a time of terror. Rather, the assassinations usually appeared in the West as psychological aberrations; or traditional violence of distant and underdeveloped societies; or, as in the case of De Gaulle, the result of special and temporary conditions.

The first perception of a "new" direction in assassination came with attacks on airliners on the ground, where the targets, despite protests to the contrary, were clearly passengers, guilty of no more than being in the wrong national airliner or of being citizens of a "guilty" nation. Hijacking and hostage-taking were bad enough—with their implication that the hostage's life, not simply his freedom, was a matter of negotiation—but firing machine guns into planes

[2] All that is except for the Puerto Ricans (quite rational fanatics) who tried to kill Truman.

filled with innocent passengers was an even more desperate matter.

At the same time, in Latin America a second assassination variant appeared: the killing of diplomats. On January 16, 1968, in Guatemala City, gunmen in a passing car shot to death Colonel John D. Webber, commander of the thirty-five man United States defense attache's office, and Lieutenant-commander Ernest A. Munro, head of the group's naval section. Two enlisted men were wounded. The next day the Revolutionary Armed Forces (FAR) claimed credit for what appeared to be a failed kidnap attempt. On August 28, United States Ambassador to Guatemala John G. Mein was shot to death during a kidnap attempt. Thus traditionally protected foreign officials had become revolutionary targets. Any "imperialist agent" could and soon did become a victim. In 1969 Charles Burke Elbrick, the United States Ambassador to Brazil, was kidnapped by four men who ambushed his limousine in a street not far from the embassy in Rio de Janeiro. The Revolutionary Movement-8 (MR-8) demanded the release of fifteen prisoners and their safe conduct to Algeria, Mexico, or Chile within forty-eight hours or Elbrick would be executed. On September 5, the Brazilian government agreed to the demands, broadcast the MR-8 manifesto, and the next day flew the fifteen prisoners out to Mexico. Two days later Elbrick was released. Diplomatic kidnappings became the wave of the revolutionary future. In 1970 there were seventeen incidents, mainly in Latin America, involving a variety of nationals, including Russians and Japanese. Diplomats had become targets—and victims—seized as symbols of terrorist-declared enemies, manipulated for real and imagined gain, shot or released in no discernible pattern. Diplomatic immunity had become an anachronism. The representatives of imperialism were "guilty," just as were those who flew on the wrong airlines or held the wrong passport.

By 1970 some of the threatened began to suspect that the hijackings, random massacres, and the assassination of foreign officials might represent an entirely novel wave of revolutionary violence. Yet in the previous decade in Venezuela nearly all these "novel" techniques of terror had been employed.

In 1962, after repeated guerrilla failure, the *Movimiento de Izquierda Revolucionaria* (MIR), radical idealists, and the disgruntled officers-turned-guerrillas set up the *Fuerzas Armadas de Liberación Nacional,* (FALN) in an effort to overthrow President Romulo Betancourt. Popular protests in the streets had been repressed, an insurrection had not been provoked, and a rural guerrilla campaign had foundered. There had to be another revolutionary road to power—and the FALN fashioned the mix of techniques and tactics that would take the revolution into the cities and onto the front pages.

First, the FALN not only shifted the area of operations but also the focus—targets were symbolic as well as functional. In October 1962, FALN guerrillas began attacking American targets: the American Creole Petroleum Company, American oil company pipelines, a Sears and Roebuck warehouse, and the United States Military Mission headquarters. Coupled with the moves against imperialist targets were stunts that attracted wide interest, amused the public, and embarrassed the government. On January 16, 1963, an FALN unit raided an exhibition of French paintings, carrying off works by Cézanne, Van Gogh, Gauguin, Braque, and Picasso. All were returned three days later—amid great public interest. On February 11, the cargo ship *Anzoategue* was seized by armed stowaways who radioed revolutionary messages back to a fascinated Venezuelan audience. On August 24, FALN "policemen" kidnapped an Argentine football star; he was released two days later. The FALN then kidnapped the deputy chief of the United States Military Mission and released him eight days later, with shoe polish in his hair. The next day they hijacked an airliner, dropped leaflets, and forced the pilot to fly to Trinidad. Simultaneously with the Robin

Hood tactics, the FALN opened a far more deadly attack on the Venezuelan security forces. The cities became guerrilla zones and the FALN killed at least one member of the security forces each day for five hundred straight days.

The FALN, with a mix of the spectacular and the lethal, sought to destroy civil order and then, with the aid of the revitalized rural guerrilla columns, to sweep into power over the ruins of a discredited regime. Finally, however, on December 1, ninety percent of the electorate went to the polls and elected *Acción Democrática* candidate Raul Leoni president. The people had voted for peace and quiet and an end to spectacular violence—and the FALN's campaign subsequently dwindled away and was largely forgotten. Ten years later, journalists and security experts and scholars—ahistorical almost to a man—would talk of the "brand new," and "unique," phenomenon of terrorism, with its hijackings, kidnappings, spectacular stunts and retaliatory raids, its assassination squads and urban guerrillas. What *was* novel after 1970 was the perception of the threatened—the realization of the vulnerability of open, postindustrial societies, and the impact of instant communications.

The Revolutionary Shift: Spectacular Violence

By 1970 a confluence of certain revolutionary forms and postures and the rise in the level of visibility of certain techniques had a more general impact. Where before the "new" terrorism had been essentially isolated—in Venezuela or in some other singularly parochial combat zone—by 1970 both the observers and the practitioners attracted a greater and more interested audience. Terror had become trendy, as a means of political action, as a threat to order, as a subject worthy of analysis and, most of all, as live drama. It was instantly transmitted in living color on the evening news.

The implications were not lost on the revolutionaries. In the postwar world the most effective, and practically the only effective, revolutionary campaigns—except for Castro's victory in Cuba—were against overt imperialist powers.

After Cuba, however, there were few revolutionary triumphs, not even over colonial powers; the Portuguese held their own in Africa, and the Americans were taking up the slack in Southeast Asia. The future appeared uncertain. In Latin America every revolutionary attempt had failed; and in 1967 Guevara had been killed in the last, most futile effort of all. In Asia Ho and Giap had not won—although certainly they had not lost—and there was no end in sight, while elsewhere, in Ceylon, Indonesia, Thailand, Burma, and India, there were no signs of a revolutionary surge. In Africa the winds of change seemed to be blowing from south to north: the South African nationalists had been repressed or driven into exile along with their South-West African colleagues. The opponents of Rhodesia had been unable to wage an effective guerrilla campaign, while the entire antiimperialist, Arab world had been devastated by the June 1967 war.

Most of the old revolutionary heroes were gone or had become paragons of stability. Nehru had died in 1964. In Indonesia, the increasingly corrupt Sukarno had been removed in 1966. The megalomania of Nkrumah led to a successful coup in Ghana and saw the liberator into exile. Che was dead and Castro was seemingly more interested in sugar production than the export of revolution. The strategies and tactics of the old heroes, if splendid, now seemed irrelevant. A new generation wanted a new direction. The New Left saw an increasingly rigid world system of international imperialist-capitalism—brutal, effective, and repressive. There was no hope of a cataclysmic solution—a spontaneous mass revolt or a world war. The Russians and, increasingly, the Chinese had grown conservative under the nuclear shadow. A few revived old heroes—Leon Trotsky and the obscure Fourth International—but for the many there had to be new heroes. They found them in themselves.

Some in the New Left were content to debate and publish endless obtuse analyses. Others came for the action, finding for the moment a natural niche on the barricade. A few chose the bomb. In America the Weathermen splinter group of the Students for a Democratic Society went underground and began the construction of bombs. In England it was the Angry Brigade. In Germany the *Rote Armee Fraktion* (Red Army Faction) of Andreas Baader and Ulrike Meinhof went beyond arson to murder. In Italy there were the *Brigate Rosse.* All were ideological kin, tiny groups of ideologues of the New Left who felt that someone somewhere must begin an armed struggle. And for many, more cautious, less committed, more stable, the bombers were romantics, their motives pure, their tactics spectacular, only their strategy in doubt.

The ultramilitants of the New Left insisted on their alliance with the revolutionaries of the Third World. They shared the same dreams and thus employed the same tactics, appropriate or not, against a single enemy. This imperialist-capitalist-racist enemy—the system—was seen as most brutal in what might be considered a revolutionary's enemies list: American imperialism, especially in Vietnam; the racism of Rhodesia and South Africa; the colonialism of Portugal; the Zionist presence in Palestine; the capitalist pawns in South Korea and Taiwan; and any conservative regime. The hero regimes were part of the socialist camp, mostly the Third World freedom fighters and especially the Vietnamese. With few exceptions the New Left in the postindustrial countries and the alphabetic freedom fighters—MPLA, SWAPO, FLN, NLF, ANC—shared not only these postures but also military weakness. In order to transform the weak into victors, the revolutionaries often chose to shift targets, from the security forces to more vulnerable sites or victims. Those who were doing fairly well in the bush, like the *Frente de Libertação de Moçambique* (FRELIMO) saw no need for such a shift, suspected spectaculars, and eschewed terror in order to polish an image as a counterstate. It was those who could only define their opponent and not maim him who chose terror. For the feeble and frustrated, terror was the only option in the new ideological garb.

The most spectacular example of just how significant the fervent few might be occurred in 1970. On September 6, the PFLP had orchestrated a vast, simultaneous hijacking over Europe—a Pan Am 747, a Swissair DC-8, and a TWA 707. An attempt on an El Al flight failed: Leila Khaled, who had taken over the TWA flight out of Rome the year before, was captured; Patrick Arguello, one of the new revolutionary mercenaries who could be hired with a slogan, was killed; and a steward was badly wounded. In the meantime the PFLP directed the Swissair and TWA flights to an old, unused British airfield near Amman in Jordan—Dawson Field. The Pan Am 747 landed at Beirut and was wired for explosives, then ordered to Cairo Airport, where the passengers were removed and the plane destroyed. On September 9, a BOAC VC-10 on route from Bombay to London was hijacked and directed to Dawson Field.

The three planes remained on the ground while the PFLP made demands that various revolutionary prisoners be released. There was a tense atmosphere of muddle and threat. On September 12, the passengers were removed to Palestinian refugee camps in Amman and the three planes blown up—in front of a worldwide audience brought live and up-close by means of television and the communications satellites. King Hussein moved in his army to crush the fedayeen. The hostages were released in clumps amid the confusion and in response to concessions abroad. The Jordanian army shelled the refugee camps and mopped up the fedayeen. At the very end of what the Palestinians would call Black September, President Nasser managed a face-saving truce. That same evening he collapsed; he died the following day.

It had been an incredible sequence of events. Buried in the Pentagon in Washington is a group known as SAGA—Studies, Analysis and Gaming Agency—whose members devise scenarios to play out war games in simulated crises. On the wall is a plaque listing the September events from the failed attempt on the El Al plane to the death of Nasser. It was a scenario that not even the most flexible military mind could have imagined creating for a SAGA game. And in response to the Jordanian repression, Black September was organized to avenge the wrong done to the Palestinian people.

For the first time the September events indicated that there might be a global terrorism problem. On October 4, 1970, the *New York Times* noted that the "guerrillas seem to have put together the techniques of the Latin American kidnappers and the Fidelista airplane hijackers." President Nixon announced that the hijackings had "brought the world to an awareness of the fragility of the network of international air traffic." Others pointed out the murder of diplomats, the rise in assassination, the struggle for the cities by the urban guerrilla. In Northern Ireland in 1971, the IRA had opened a campaign of sniping, bombing, and arson. In Spain the Basque ETA had begun to create serious turmoil. In France the Bretons turned to the bomb. On December 22, 1971, the Baader-Meinhof group killed a policeman during a bank raid. There had been bombs in Italy for two years, detonated both by the far Left and the far Right. A time-bomb exploded at a hotel used by the United States Air Force in Athens. In April the Croatians shot the Yugoslavian ambassador to Sweden. All this in tranquil, democratic, postindustrial Europe. And beyond Europe, there were kidnappings, hijackings, assassinations, the spoors of the new terrorists who, as the Tupamaros, waged urban guerrilla war in Uruguay or, as Turkish freedom fighters kidnapped United States airmen near Ankara. And the threatened, and so many seemed to be, wanted explanations and reassurances. Who were the new gunmen and guerrillas? What did they want? Why did they resort to such appalling methods? What could be done? What should be done? They did not want learned explanations of the roots of contemporary violence. Whether the experts saw the violence as new or old, anyone who could read or watch television could see that something had gone wrong. And those responsible, in Washington or Rome or London, thus had to respond to a pheonomenon that few understood. No matter, the vulnerabilities of the international system had become apparent, the threat was real. Safeguards had to be found, a deterrence system fashioned. *Something* had to be done. And as is often the case in crisis management, each response seemed *ad hoc*—a mix of national prejudices, political considerations, and growing frustration.

PART TWO

The Response: Uncertain Remedies to Real Threats

Preserve me from the violent man.
Psalm 40

Life for life, eye for eye, tooth for tooth, hand for hand, foot for foot, burning for burning, wound for wound, stripe for stripe.
Genesis

To everything there is a season . . . a time to kill, and a time to heal . . . a time to keep silent, and a time to speak. . . . a time of war, and a time of peace.
Ecclesiastes

Blessed are the meek: for they shall inherit the earth.
Matthew

Although the drama of the hijackings and the Dawson Field confrontation that led to the formation of Black September indicated that there was a new terrorism and a new threat to international order, it was not immediately clear that everyone had a problem or that anyone could be a victim. The inclination was to see such political violence as a special case, depending on the vantage point. The Americans' problem was the hijackings to Cuba; the British had, as always, the rebellious Irish; the Israelis, to meet the new direction of the Arabs' long war campaign; Sweden or Australia, to deal with truculent aliens. By the time, five years later, that the Croatians took over TWA Flight 355, there was wide agreement that a general problem existed, more serious in open societies but not unknown in Communist Europe, at least in the form of hijacking. Even democratic countries without nationality problems and without severe domestic discontent became arenas where the international terrorists staged operations or the television revolutionaries choreographed spectaculars. So everyone was threatened and anyone could be involved in a novel variant of political violence that was still, even in 1977, little understood.

In the first round of response to the new terrorist challenge in 1972–1973—Lod-Munich-Khartoum—most of those responsible for safeguarding innocents or negotiating with the terrorists acted only with the most limited guidelines, under great pressure, with inadequate intelligence, dependent upon their own predilections, with uncertain aid and comfort from their governments. Simultaneously, especially after the Munich massacre, a concentrated effort, particularly in America, was undertaken to analyze the terrorist phenomena, to find a consensus on causes and effects, and most of all, of course, an appropriate response. This search for a common wisdom produced mixed results. Most of those involved produced congenial ideological or methodological findings, usually too ethereal for policy makers and often buttressing long-held political positions.

If the analytical effort, particularly by academics, left much to be desired, the West's vulnerability to terror was generally accepted. But how serious and broad the threat, how real the vulnerabilities, and how necessary increased safeguards remained matters of considerable dispute. Some felt that the terrorist threat was transient, others did not. The evidence of the vulnerabilities, however, was undeniable, and there was very real disagreement as to what was to be done to safeguard societies. Simultaneously the public perception of an escalating threat grew, and with it an outraged indignation that the new terrorists seemed to act with impunity, to evade punishment, and to breed global disorder while the responsible did little or nothing. There were concessions and deals and still more violent incidents. *Something* had to be done.

Actually, a great deal was being done, not in broad analytical terms nor in fashioning a general Western, or even national, strategic response to the problem but in the creation of technologies and tactics that safeguarded vulberabilities, and in a halting but significant attempt to construct an international (or at least Western) consensus on appropriate legal remedies. Not unexpectedly, it proved easier to search airline passengers than to achieve a United Nations covenant; but both directions, the technical-tactical and the international-legal, eased public anxiety, protected democratic institutions and liberties that the desperate often seemed to want to discard in the cause of vengeance or in the name of security.

Still, in the four years after the Khartoum incident in 1973, most of the same problems remained, although governments did have a far greater range of technical and tactical assets and consequently more options. Very few Western governments managed a coherent strategic response but generally opted for *ad hoc* responses within certain gradually accepted guidelines. Some were aggressive and adamant, others flexible, and a few swift to concede. There was increasing agreement on legal measures, on international covenants, on Western security cooperation. But each new violent deed tended to occur in a novel context where past experience often proved a faulty guide and the law of limited application. If the Western response was halting and the remedies often uncertain, still the threat was accepted as real but not because of the numbers involved or the butcher's bill. The crucial difficulty was in balancing an overreaction that would erode Western liberties against an insufficient response that would erode public morale and, consequently, equally endanger Western democratic institutions. At least by 1977, an increasing number of those involved in policy decisions recognized that this was the vital dilemma and that there was no single, simple solution. In essence the problem was that there was no solution.

CHAPTER

NATIONAL POLICIES: THE FIRST ROUND, 1970–1973

. . . we cannot have governments—small or large—give in to international blackmail by terrorist groups.

President Richard Nixon,
October 1973

It was widely agreed that the problem of revolutionary terrorism was peculiarly challenging to an open, democratic society. Brutal, efficient, authoritarian governments had to cope with occasional hijacking or bomb incidents, but no "terrorist problems"; however, democratic governments could hardly remain open and democratic by opting for repression. And even when a democratic government was capable of accommodating internal pressure for radical change and had no unresolved nationality problem, international or transnational terrorists could use the country as a stage. The Croatians, beginning in New York, appeared in Canada, Iceland, over Britain, and ended in France. Thus, all open societies were vulnerable, no matter how stringent the safeguards.

In 1972 and 1973, as a result of Palestinian operations in the Middle East, Europe, and Western Europe, it became clear to Western governments that the new terrorism was a general rather than regional or nation-bound threat. It seemed that a new generation of revolutionaries had arisen without restraint or remorse, bound together in an international conspiracy to disrupt global order. While it was obvious that Israel had become the prime terrorist target before 1971, increasingly the Palestinians and their ideological allies everywhere in the West insisted that the imperialist-capitalist-racist system was the enemy; that is, all open societies were fair game. The incidents at Lod-Munich-Khartoum were the watershed in Western perceptions. Spectacular tactics involving innocent victims had previously been accepted as novel revolutionary techniques, but in special campaigns—not as evidence of a general assault on the West. After Lod-Munich-Khartoum, to a greater or lesser degree, every democratic government had a terrorist problem—in each case a slightly different problem but still, demonstrably a terrorist problem.

In 1972 the most experienced of the threatened governments was Israel, a special case. Surrounded by Arab states that denied its legitimacy, Israel from the first had never been really secure. For Israel there had always been an emergency, at best a state of no-war-no-peace. Curiously, however, the transformation of the nature of the Arab threat to Israel came after the June 1967 war when secure boundaries had been achieved, when formal military might of the Arabs lay in ruins, when Jerusalem had been reunited and a sense of euphoria filled the people. It was then, when all other means to influence events seemed lost, that the Palestinian Arabs opted for irregular war.

From the autumn of 1967 until the autumn of 1970 the Arab fedayeen, dominated by Al Fatah, undertook a guerrilla war in the occupied zones. The campaign was a failure. Israel had put together a national policy in response to the fedayeen threat: immediate harsh retaliation to any provocation. By recourse to air raids, by artillery and mortar attacks, by commando raids, Israel met force with force. Except in Israeli announcements intended for foreign consumption, there was not much pretense that the policy would deter the Arabs: retaliation was intended only to punish the terrorists and to assuage the Israeli population.

The next problem lay in punishing elusive fedayeen when by 1970 they began attacks in distant cities beyond the reach of the Israelis. At first such novel Arab tactics produced an uncertain Israeli response, as in 1968, for example, when Israel exchanged sixteen Arabs for the crew and passengers of the El Al airliner skyjacked to Algeria. But subsequently the traditional strategy of maximum response against the guilty and their patrons continued. Blaming the Lebanese for giving the fedayeen sanctuary, in December 1968, Israeli helicopter commandos hit Beirut Airport and destroyed thirteen Arab airliners. Although outraged, the Lebanese and the other Arab states became more discreet about fedayeen activities in the future.

Just as in the United States, where the hijackings had been regarded as largely an American-Cuban problem, so the cycle of provocation and retaliation in the Middle East tended to be viewed as a regional aberration arising from special, if long-lived, rivalries. Even the Israelis did not really consider the fedayeen as a novel or even especially important challenge, just a highly visible one. Their traditional policy of retaliation was simply refined and expanded following the June War and the campaign of attrition along the Suez Canal. Moreover, the erosion of the fedayeen presence by the Jordanian army in the September 1970 debacle appeared to ease Israel's security problems. With most other options denied, the fedayeen were left with terror, often in arenas far from Israel. Then in May 1972, Habash and PFLP found a way into the country on Air France Flight 132 from Paris to Rome.

Lod, Israel, 1972

Among the 115 passengers who landed at Israel's Lod Airport on May 30, 1972, were three young Japanese "tourists." After the passengers had disembarked and moved into the arrival lounge to collect their luggage from the conveyor belt, the three Japanese lifted their fiberglass suitcases from the conveyor, took off their jackets, opened the cases, and stood up holding shrapnel grenades and Czech VZ-58 automatic assault rifles without stocks. Almost simultaneously the three began to sweep the lounge with automatic fire, pausing only to toss out the grenades. The Japanese Red Army (*Rengo Sekigun*)—in alliance with Habash's PFLP—had arrived.

In an instant there was bloody chaos: the chattering guns, the shattering crash of grenades, the screams, the smell of cordite and soon of blood. The floor was strewn with baggage, torn and bleeding bodies, passengers scrambling and cowering. Some tried to reach the security of restrooms or the first-class lounge. Others crouched behind chairs. Israeli security people in the turmoil could find no targets. Professor Aharon Katchalsky, a leading authority on biophysics, lay dead. Then a wild burst hit and killed one of the terrorists—Yasuiki Yashuda. A few seconds later, Takeshi Okidoro held onto a grenade a moment too long. It detonated and blew off his head. The single survivor, Kozo Okamoto, his VZ-58 empty and his first two grenades gone, rushed from the lounge onto the tarmac outside the terminal. He was tackled by an El Al traffic officer, Hannan Claude Zeiton, and held down until reinforcements arrived.

Gradually the Israeli authorities put together a report of *what* had happened; finding reasons for the attack was somewhat more difficult. The three Japanese of *Rengo Sekigun* had traveled from Tokyo to Beirut and back to Rome, where they had boarded the flight to Lod. (Their luggage had not been searched, although there had been a body search at the gate.) As allies of the PFLP, they had undertaken their kamikaze mission because "The Arab world lacks spiritual fervor, so we felt that through this attempt we could stir up the Arab world." Defending the action, Ghasson Kanafani, editor of the PFLP's *Al Hadaf,* told reporters, "Our style of operation is not an invention of a person but a result of our situation. If we could liberate Palestine by standing on the borders of

South Lebanon and throwing roses on the Israelis we would do it. It is nicer. But I don't think it would work."[1] He dismissed all criticism of the Lod incident as "unscientific, unmoral reactions that condemn only us and ignore other tragedies, especially those committed by Israel." As for Okamoto's summation, it confounded rational opinion:

> We three soldiers, after we die, want to become three stars of Orion. When we were young we were told that if we died we may become stars in the sky. I may not have fully believed it but I was ready to. I believe some of those we slaughtered have become stars in the sky. The revolution will go on and there will be many more stars. But if we recognize that we go to Heaven, we can have peace.[2]

Some Israelis found comfort in considering the Lod massacre in the same category as a natural calamity—like a hurricane or a flash flood—beyond rational preparation. Most, including the government, believed that negotiations or accommodations with the likes of the Okamoto-Habash fanatics was a waste of time and counterproductive. The basic axiom of the Israeli terrorist policy—no concessions ever—remained unquestioned. The appropriate response to provocation would be retaliation. On July 9, Kanafani, who had dismissed all criticism of the Lod incident, and his seventeen-year-old niece Lamis Najem were killed by an exhaust-pipe bomb that detonated when he turned on the ignition of his sports car.

Two aspects of the Lod massacre were especially disturbing to those not intimately concerned in Middle East matters. First, the cooperation of the Japanese Red Army and the PFLP reinforced suspicions that a terrorist-international was evolving. There had been regular, if not very authoritative, reports of links between the Arabs and the IRA or the Turks or the German left. The three Japanese, for example, had named their mission after Patrick Arguello, the South American killed beside Leila Khaled during the attempt on the El Al

[1] *Washington Post,* 9 July 1972.
[2] *New York Times,* 14 July 1972.

flight in September 1970. Such a conspiracy posed serious international problems; especially if, as seemed likely, there were patron regimes (Libya being the prime suspect) for such operations.

Perhaps even more disturbing, the PFLP announcement after Lod that tourists, whatever their nationality, were enemies, reinforced fears that the new terrorists were becoming increasingly indiscriminate. The targets were no longer simply Zionists—any Zionist, any age, any place—or their own Arab traitors, but anyone remotely connected with Israel and also, according to Okamoto, ". . . we will slay anyone who stands on the side of the bourgeoisie."[3] The list of potential revolutionary targets seemed endless—and so did the problems of security.

Israeli authorities took the most rigorous measures to reduce the vulnerabilities of facilities abroad and of El Al. Israeli embassies and consulates were supplied with armed guards, walls and windows were reinforced, closed-circuit television cameras were installed, rigorous inspection systems for parcels and visitors were introduced. Diplomatic personnel were constantly made aware of the potential dangers the fedayeen represented. Arab letter bombs, assassins, car bombs, commando raids, and diabolic devices were all possible, and sooner or later all were employed. There was little that was elegant or graceful about the Israeli embassies, but penetration was no easy matter. El Al, too, developed most detailed precautions. All luggage was inspected, not just carry-on bags. Inspectors typed on typewriters, sprayed aerosol cans, took tape recorders apart. Specially trained dogs were on hand to smell out plastique. Those who had memorized photographs of known fedayeen stood unobtrusively in most of the worlds' airports. There were discreet armed guards in the airport and inside the jets, with orders to shoot if need be. There were other precautions

[3] Ibid.

on landing. And the death of Kanafani on July 9 indicated that Israeli policy was not wholly defensive, even if his death did not indicate a full-fledged campaign of counterterror.

Munich: Germany, Israel, and the United States

If the Israelis, long in a state of no-war-no-peace, believed in "excessive" preventive measures, others were under no such compulsion. They were not engaged in a low-intensity war. They had no special domestic enemies. They felt no overpowering urgency to strike back at every provocation. They were, in fact, not involved—or rather, they hoped that they would not be.

But that same year, on September 5, the illusion of innocence and isolation was shattered for more than one government. At four o'clock in the morning, eight Black September fedayeen in track suits, carrying large bags filled with arms, climbed over the fence into the athletes' quarters of the Munich Olympics. They moved directly to the Israeli dormitories, seized nine hostages and in the process killed two other Israelis, Joseph Romano, a weightlifter, and Moshe Weinberg, the wrestling coach. By dawn the Munich Olympics had become a gigantic stage for another terrorist spectacular. Practically every television organization in the world was present. And the cameras were connected to most of the globe by satellite, so this real-life drama would be enacted for untold millions of viewers.

The fedayeen presented a list of demands. Israel must comply or the hostages would be killed. Israel refused and urged German Chancellor Willy Brandt not to give in to blackmail. All day negotiations continued, as one deadline after another passed. Brandt tried to get the Egyptians to accept the hostages and fedayeen in Cairo. The Germans felt that the Israelis could then be returned and the Black September fedayeen released. Egyptian Prime Minister Sidki declined the proposal. Egypt did not want to be involved. Finally, the Germans decided to pretend to concede to the ultimate demands and provide the fedayeen with helicopters to fly the fedayeen and the hostages to the airport, where they would be provided with a Lufthansa airliner that would fly them on to Tunis.

The two helicopters were moved up close to the quarters. The fedayeen, glued to their hostages, were flown to Fürstenfeldbruck military airfield. The Lufthansa Boeing 707 was parked, ready for inspection. Two fedayeen walked across the tarmac and inspected the empty jet. On their way back the Germans opened fire. It was the beginning of a ninety-minute battle that cost the lives of all nine hostages (killed by the Arabs), five of the fedayeen, and a German policeman. One of the volunteer helicopter pilots was seriously wounded. It had been a debacle. In face of the global wave of revulsion and outrage, the fedayeen spokesmen were unrepentent as usual:

> They never cared about us. Why should we care about them? Call us what you may, but it is good for our morale, and it may help the moderate elements in the movement to take a more militant position. After all our defeats, this comes as an uplift. We feel we have to do something. What does the world expect of young Arabs these days? We have seen too many defeats.[4]

Germany

Although Premier Golda Meir thanked the West German government for their efforts, there were swift changes as a result of Munich. The German government set up a special antiterrorist group of marksmen, commandos, and technical experts. Extensive and intensive measures were taken in regard to the large number of Arab workers and students in Germany—the key inside man at Munich had been an Arab

[4] Ibid., 20 September 1972.

architect. These were essentially tactical, or in some cases technical, responses. There were much more revealing strategic shifts as events seven weeks later showed.

On October 29, two members of Black September hijacked Lufthansa Flight 615 from Damascus to Frankfurt. They held it and the passengers for ransom until the three survivors of the Munich operation were released to them at the Zagreb Airport. The German government agreed with such alacrity that many suspected the "hijacking" of Flight 615 had been previously arranged. In any case the three members of Black September arrived at Zagreb and together with their two comrades flew off to Libya, increasingly used as a terrorists' sanctuary.

Like other stage-nations the Germans did not want to hold "poisoned pawns" who would engender further rounds of terrorist hostage-operations. Unofficial estimates are that in the next five years 141 of the 150 Arab terrorists arrested in Western Europe were released without trial. After all, Germany had been the *site* of other peoples' wars, not the target—except, accidentally, for the policeman victim. Even the Olympics had been chosen not because the games were in Munich but so that the telecasts would underscore that unless there were justice for Palestine, there would be no peace for anyone, not even at the Olympics—perhaps especially at the Olympics.

It was not easy, however, to deal with native terrorists who could not be deported to some distant sanctuary. During the previous two years Germany had been plagued by the small group of radicals, the *Rote Armee Fraktion* of Andreas Baader and Ulrike Meinhof, who launched a series of sabotage operations. In 1968 Baader was convicted of arson, disappeared while on bail, was rearrested in Berlin, and then freed in May 1970 by Ulrike Meinhof and others during an armed raid. The Baader-Meinhof group became a television and press sensation, but the crucial moment came in May 1972 when United States military facilities were twice bombed and four Americans were killed. In June 1972, Baader, Holger Meins, and Jan-Carl Raspe were arrested in Frankfurt, and Ulrike Meinhoff was picked up in Hanover. It soon became evident that their friends, some in the similar *Bewegung 2 Juni* (2 June Movement), were still operative, that contact in and out of prison was possible, and more alarming, that they had intimate contacts with the fedayeen. In 1973 RAF published *Die Aktion des schwarzen September in München* praising the Munich massacre by Black September. German authorities decided to build a special prison-court to isolate the RAF people from outside contact—at a cost of twelve million Deutschmarks—and to seek legislation limiting the number of defense attorneys. Neither tactic seemed to limit the operations of RAF and the allied Second of June Movement. There was growing concern about how to fashion an appropriate response in a society still fearful of a heritage of repression.

Israel

Still the most threatened of all, the Israelis took certain remedial steps: three senior members of the Israeli Department of Internal Security were dismissed, and further efforts were made to tighten all official security at Israeli embassies and consulates, as well as on El Al flights. Golda Meir appointed Major General Aharon Yariv as director of *Mivtzan Elohim* (The Wrath of God) to direct an antiterrorist campaign employing the techniques and tactics of terror in any arena. Leaders and agents of Black September, the PFLP, and other fedayeen organizations became prime targets—victims of car bombs, assassination squads and explosive devices in telephones in Rome or Madrid or Nicosia. On April 10, 1973, an Israeli team went ashore near Beirut to meet six agents-in-place. Two and a half hours later, the attack squads returned. Three Al Fatah leaders—Abu Youssef, Kamal Adwan, and Kamal Nasser—had been killed, along with Abu Youssef's wife and bodyguards and an innocent spectator. Other diversionary attacks were made on fedayeen camps and headquarters. In Israel General David Elazar told a press conference that the terrorists' leaders should realize Israel's capacity to fight everywhere.

The United States

Although no Americans had been involved in the Munich massacre, the events of September 1972 had a profound impact in Washington. As leader of a law-and-order administration, President Nixon saw Munich as a direct challenge to world order, a most serious provocation. He announced the formation of the Inter-Department Working Group on Terrorism, chaired by Secretary of State Henry Kissinger. Located in the State Department and initially directed by Armin Meyer, the former Ambassador to Lebanon, the board consisted of representatives from State; the U.S. mission to the United Nations; the Central Intelligence Agency; the Federal Bureau of Investigation; Defense, Treasury, Transportation, and other relevant departments. The mission of the group was to coordinate governmental policy and intelligence, cooperate with concerned countries and regional organizations like NATO and CEATO, consider tactics, and in an emergency set up task forces.

But in the deepening morass of Watergate, Kissinger had little time for a Cabinet committee on terror, and most of the group members had little leverage, a most limited brief, and no real budget. The group was in fact a traditional American response to political trauma, a commission to ease public anxiety. It was, of course, seen by Arab observers as an institutionalized step to join in Israel's antiterrorist campaign rather than a well-meaning palliative for American outrage and frustration.

Elsewhere, especially in Europe, the tendency of the threatened governments was to concede or to punish the terrorists under normal statutes and then, with varying rationales, to release them in order to prevent still further incidents. America's unusual hard-line stand, the establishment of the Inter-Department Group, and the President's public admonitions only provoked the frustrated fedayeen. They assumed they faced an open collusion between Israel and the United States to murder the Palestinian leadership. In turn the Voice of Palestine broadcasts threatened "death to the Americans."

The United States: Khartoum

There were many in Black September and Al Fatah who actually preferred to strike at America. They accepted the fact that Israel's stringent security arrangements coupled with her refusal to bargain, much less offer concessions, made manipulative terror operations, or any operations, difficult. American targets might be preferable. So for a variety of reasons they selected a diplomatic reception in honor of an American at the Saudi Arabian Embassy in Khartoum. United States *Chargé d'affairs* Curt Moore, a highly popular envoy was about to be replaced by the new American Ambassador, Cleo Noel.

On the evening of March 1, 1973, Black September raiders rushed into the embassy waving Kalashnikov assault rifles. Some guests departed over the garden wall, others were allowed to leave, but Curt Moore, Ambassador Cleo Noel, Belgian *Chargé* Guy Eid, the Saudi ambassador, and the Jordanian *Chargé d'affaires* were herded together as hostages. The Black September spokesman demanded that the Jordanians release seventeen members of Black September, including Abu Daoud, suspected of organizing the Munich massacre the previous September; that the Americans free Sirhan Sirhan, the convicted killer of Senator Robert Kennedy; that the Israelis release Arab women fedayeen prisoners; and that the Germans release the Baader-Meinhof prisoners. Under a glowering sky, as sandstorms began moving into Khartoum, the negotiations dragged on with shifting demands but no concessions. The Jordanians would not give in. And in Washington President Nixon declared publicly before the television cameras that the United States "cannot and will not pay blackmail."

Black September could hardly lose. Either the imperialists were coerced into making concessions and the hostages released, or they refused and the hostages would be killed, simultaneously emphasizing the reality of future threats. There were some in the organization who preferred the latter, deadly alternative. Outside the Saudi embassy, standing in the sandstorm, the security forces and the correspondents waited in the dark. It was clear that no concessions would come out of Washington or Amman. At a little after nine in the evening, they heard a muffled burst of shots. Moore, Noel, and Eid had been murdered. A few hours later the fedayeen surrendered. They were tried, convicted, and sentenced to death; but the sentence was commuted, they were released into the custody of the Palestinian Liberation Organization and flown to Cairo. El-Nimeiry did not want the fedayeen in his prison in Khartoum any more than the Germans wanted their Black September prisoners in Munich. President Nixon announced that Khartoum underscored the need for a firm international stand against the menace of terrorism. Nixon chose the response of no-negotiation, no-concession, thus leaning toward the policy of Israel and away from those who were more concerned with hostage safety.

Equally significant, however, was the American administration's realization that the new terrorism demanded a reasoned response. After Munich and the murders at Khartoum, an increasing emphasis was placed on devising an American strategic response to the new terrorism. And in one area at least there was progress. Thus to a very considerable extent the problem of hijacking was on the way to solution as a result of a mix of diplomacy, research, and improved security. In 1973 a bilateral agreement with Cuba ended sanctuary. Psychological research produced an effective profile of potential hijackers that helped security attendants to filter out nearly all the psychotics. And improved security—searches, tarmac protection, and, for a time, air marshals—placed severe obstacles in the path of criminals or the demented.

The same mix, it was hoped, might prove effective against international terrorism. A formal effort was undertaken immediately to achieve United Nations consensus on the problem but with no success. So, where possible, initiatives were undertaken with sympathetic governments. New physical security precautions were taken at all American diplomatic facilities, new visitor checks were established, new devices and doors installed, and the Marine guards were no longer considered ornamental. Additional funds were invested in protecting foreign embassies and diplomats in New York and Washington. And the policy of no-negotiation, no-concession was maintained, despite criticism that it was a posture and not a policy that under certain circumstances might have to be altered.

Still, after Munich it became increasingly apparent that there were real difficulties in responding to the new dramas that were soon mimicked in whole or in part by various revolutionary groups and by those on the fringe of political reality, such as the Hanafi sect in Washington. Matters were not as simple as Nixon's no-blackmail posture indicated. Some states, apparently, faced no real threat. Brutal, efficient authoritarian states appeared largely immune. One Budapest 1956, or one Prague spring every ten years seemed to be the rule. The rebel would choose violence only if there was some hope of later success; he would remain quiet, albeit grudgingly, if there was no hope. But the world, especially the Third and Fourth World, was filled with less efficient but often brutal authoritarian states—a dangerous combination for rulers of questioned legitimacy. Some regimes were sufficiently fragile that the obvious route to power was an assault on the presidential palace rather than a terror campaign. And by the 1970s only a few tiny rocks and enclaves were left from the old, vulnerable empires. Increasingly, it was open societies with responsive governments operating under the rule of law that were the targets of the terrorists.

The problem was how to respond effectively to indigenous or imported violence within democracy. Recourse to brutality, efficient or not, appeared to be a foreclosed option. An authoritarian Brazil could crush the urban guerrillas of Carlos Marighella by brutal repression, institutionalized torture, tolerated vigilante operations, and the suspension of any vestigal civil liberties. Uruguay, on the other hand, Latin America's showcase democracy, beset by the Tupamaros, a wealth of domestic difficulties, and the accumulated cost of past policy errors, tried to cope within the law. When the army lost patience with the government, the open society was closed, civil liberties discarded, and the methods of effective brutality initiated. Order was restored beyond the law by means that older and somewhat more secure democratic states hoped would be unnecessary. Even nations with serious ethnic problems (Canada with Quebec or Britain with Ireland, for example) or with indigenous bombers and gunmen (for instance, the Baader-Meinhof group in Germany or the *Brigate Rosse* in Italy) felt that there must be an appropriate *democratic* response.

Yet after Lod-Munich-Khartoum there was absolutely no consensus on a single democratic response; indeed, in some cases there was still considerable reluctance to accept the fact that terrorism had come to stay. The Italians recognized that there were Fascist fanatics and New Left fanatics, but these operated within an Italian context. The British had for centuries lived with the rebellious Irish and even IRA bombs in London were no novelty. Everyone agreed that the Israelis had their own special problems and their response—hard, adamant, thorough, and aggressive—would not work elsewhere. What nearly everyone in a policy position wanted and, after Lod-Munich-Khartoum, wanted with growing urgency, was information, intelligence, analysis about the "new" terrorists. Were these people rational political fanatics or crazies? Was this the wave of the future? Could there be accommodations and concessions? Was Nixon right? Were the Israelis right? It was not only a case of what is to be done but also what was going on. Until there were some rational explanations coupled with prescriptions, democratic societies were fighting a holding action founded on little more than indignation and anguish.

CHAPTER

INTERNATIONAL ANALYSIS: THE COMMON WISDOM

If ignorant both of your enemy and of yourself, you are certain in every battle to be in peril.

Art of War,
Sun-tzu

After Munich, terror became trendy—following the same general pattern as insurgency, urban riots, guerrilla activities, and American violence. Instant specialists emerged from the universities and think tanks. While "terror" has hardly had the same impact on the American academic community as Black Studies or Women's Studies, governmental concern and the media coverage did inspire considerable activity and in time created an academic-terrorist-establishment rooted in the social sciences, with a legal wing and an historical wing.

Conferences have been held within and without the government, in the United States and abroad. New journals have been launched—*Conflict and Terror* and *Counterforce.* And those in the know (or on the make) have hurried to their typewriters—the most effective weapon of the weak. Articles and then books on the subject of "terrorism" have proliferated. For terrorism *was* trendy, so that those who a few years before had toiled unnoticed in the vineyards of Middle East communications or Algerian history or the history of social movements were suddenly invited to address assembled bureaucrats or comment for *Time* and *Newsweek*.

Not unexpectedly, the flood of analysis revealed several limitations. First, except for the few historians and despite brief disclaimers to the contrary, much of the work has been ahistorical. Like the *New York Times Index*, for most writers "terrorism" did not exist before 1969. Second, virtually all the analysis was limited by the speciality of the writer. The lawyers were concerned about laws broken or laws proposed; the political scientists wanted to count, if at all possible; the psychologists wanted to generalize about the mind of the terrorist, despite the dearth of live subjects. Third, while there was no agreement on a definition, everyone was either against terrorism or an apologist for revolutionary terror. Disinterested investigators were rare. The result of this extended and partially spontaneous process has been what might be called the common wisdom. Unfortunately, in few areas are limitations and frailties of academics and their various patrons as clearly revealed as in the common wisdom concerning terrorism.

The Common Wisdom

At the very beginning we face the definitional problem: *one man's terrorist is another man's patriot.* Like love, terrorism is easy to recognize but difficult to define. Some include practically every violent act by anyone, short of self-defense as a last resort. Others can exclude even the most unsavory slaughter of innocents because of objective revolutionary conditions—that is, their own support of the killers' cause. In September 1976, the *New York Times* reported Acting Foreign Minister of Libya Abu Zoid Durda's definition of "terrorism":

> To station American forces overseas is terrorism. To monopolize the wealth of countries is terrorism. To dominate the outlets of seas and oceans is terrorism. To provide aging regimes with sophisticated weapons to oppress the people is terrorism. To use wheat and gold as political toys when the world is starving is terrorism.

He did suggest that hijacking was terrorism, and Libya stood against hijacking. (He also pointed out that Libya's antihijacking law was based on principles derived from the Koran.) Yet there is little evidence that this law has been applied, for Libya seems to be a haven for a whole spectrum of revolutionaries.

In general there is a reluctance to recognize *our own* terrorists. Americans, for example, supported the first two waves of airline hijackings, refugees from the Communist countries of Eastern Europe and then from Cuba. And it is quite simple to imagine a hijacking scenario that would still win approval: a hijacked Soviet airliner, filled with Jews fleeing Russian persecution, landing at a United States airbase in Germany. The Croatians felt they were just simply patriots—"We are defending a just cause and yet here we are with handcuffs on our wrists." And so it goes.

In the nineteenth century the word *terrorism* had a relatively clear meaning. People called themselves "terrorists" (today they are "guerrillas" or "freedom fighters" or "fedayeen") and held to a particular revolutionary strategy—personal terror, propaganda of the deed. Today the word is often used as a pejorative, so that there is no agreeable common defintion of the word.

Besides the definitional problem there is also the historical problem. Here the conclusion is that the present epidemic of terrorism is either *novel or not.* Some trace the long roots, ideological or tactical, of present day revolutionary terrorists back to the Black Hand or the People's Will or the Carbonari. To them the events of Lod-Munich-Khartoum represent simply one more turn of an old screw that merely *seems* novel. Once pirates seized ships and bandits captured trains. Now airplanes are the targets. Assassins have always been with us, and all war was once irregular.

The advocates of novelty, however, point to differences in kind, not just degree. For one thing, our postindustrial society presents a spectrum of highly vulnerable nodes. Vast computer banks, microwave relay stations, nuclear facilities—all the technological infrastructure of modern life—can be threatened by even a small, maniacal band. In 1939 and 1940 the IRA could bomb away in London with limited effect; after all the Luftwaffe would later bomb the city every day for months and yet British war production increased. Today a few dozen explosive devices in just the right places could cause very serious repercussions—nuclear fallout, strategic vulnerability, a collapse of communications. And those terrorists who seek to influence rather than to maim now have instant communication to hundreds of millions of people. Not until the 1930s could millions listen to one man, and not until the last generation could millions see live action on television. Now, with the use of communications satellites, one man shot down on camera is instantly transformed for the millions into a million murders flashing out of the box in the living room. What impact this may have remains a mystery, but most assuredly it adds a different and little-understood dimension to political violence. There seems little doubt that without the existence of television the five Croatians would still be handing out smudgy leaflets on street corners or that Ambassador Noel, Guy Eid, and Curt Moore would be alive.

One of the few firm conclusions emerging from the academic investigations is that *terrorism is the weapon of the weak, but it is a powerful weapon.* It seems clear that in most cases "terrorism" is not a desirable option for the revolutionary. Most revolutionary organizations would prefer to be counterstates employing sanctioned tactics in an irregular war. Thus terror is most often chosen at the beginning of a campaign when there are no resources but the assassin's tools (the tiny People's Will in Russia seeking to kill the Tsar) or when reverses have narrowed tactical options (Habash's PFLP turning to hijacking), or when the campaign may be about to be closed down (the Stern Group searching for a mission once Israel had been created and so murdering Count Folke Bernadotte). And as a weapon of the weak, it is generally recognized that the rationales for terrorism are the same: there is no other way and in any case we act in the name of a higher law.

What is troubling, of course, is that various contemporary vulnerabilites have made terrorism not a *weak* weapon of the weak but their best possible choice. A tremendous impact can be achieved with negligible capital or training. All that terrorists need is a willingness to take disproportionate risks with their own lives and with the lives of others.

How important is terrorism? According to the conventional wisdom, terrorism is either *very important or it is very unimportant.* On one end of the spectrum stand many of the Union Jack School, who see the terrorist threat as releasing highly undesirable forces within open societies; at the other end are those who insist that terrorism is counterproductive, ineffectual, and unimportant. Advocates of both positions, of course, are in fact engaging in special pleading, whether consciously or not. There are those who *want* terrorism to be a serious, and consequently dangerous, matter in order that a cherished response can be adopted or long-lived trends reversed. And for the same reason, there are those who want the opposite.

The Union Jack School tends to equate normalcy with the political conditions and mores of the Home Counties. Much that is new in English society—Commonwealth immigrants, the rise of violent crime, serious labor unrest, radical (Maoist or Trotskyite or Anarchist) ideas circulating among students and shop stewards—is seen as not only undesirable but also potentially dangerous. There is the fear that political violence, as demonstrated by the "symbolic" bombs of the Angry Brigade and more violently by the Provisional IRA's

English campaign, may escalate and that there may be a future need for not just riot control but antiinsurgency operations as well. The most elegant and professional treatment of the British Army's role is such an eventuality is *Low-Intensity Operations,* by Brigadier Frank Kitson; but he is far from alone in viewing the prospect with alarm. Ordinarily, but not necessarily, coupled with this fear of a fragile and vulnerable society is the assumption that the enemies of democracy and the West, especially the Soviet Union, are seeking out and exploiting such flaws, and will continue to do so. Thus the new terror is linked with the old Cold War. There are those, as well, who exaggerate terrorism in order to discourage (or encourage) other social objectives. For example, those who oppose the nuclear power industry regularly stress the prospect of sabotage or theft of nuclear materials by revolutionaries, criminals, and psychotics—no nuclear facilities, no terrorist threat.

Oddly enough, those who tend to minimize terrorism often share the same premises, stated or unstated. They too abhor this kind of violence and suspect that the "terrorists" are manipulated. But instead of stressing the dangers of terrorism, they focus on its futility. More subtly, instead of advocating a positive response to punish or destroy, they seek to deter potential terrorists. Thus Walter Laqueur, Director of the Institute of Contemporary History in London, a prolific and perceptive investigator, has regularly written on the futility of terror—"For in the final analysis, the politics of individual terror are utterly futile for the Arabs."

> In all likelihood, the nurder of innocent people will continue for years to come. But it will not continue forever, for history teaches that in every conflict—be it in Ireland, Cyprus or elsewhere—eventually a stage is reached in which even the most fanatical elements recognize the futility of their terrorist attacks. . . .
>
> Once the image of the Arab was that of a proud man, a paragon of all manly virtues; it is depressing to witness what has become of this image.[1]

[1] *International Herald Tribune,* 11 September 1972.

In other words—terror is pointless and ruins your image: so stop.

Another observer equally concerned with Arab terror is General Y. Harkabi, the former director of Israeli military intelligence. Terrorism, he suggests, is important on a tactical level but not strategically. And even on a tactical level, there are those who note that if the Palestinian problem were solved—to their satisfaction, of course—terrorism would become a rather minor international irritant. Still, others reply that whatever the underlying causes of World War I, the powder train that set the whole show in motion was lit by one man with a revolver. So it is generally agreed, for various unstated reasons, that terrorism is either very important or else it is not.

It is a comfort to know that the common wisdom has accepted as fact that, however defined, *terrorism (nonstate variant) kills relatively few people (but has a great impact).* While it is true that a terrorist deed may ultimately lead to mass slaughter as it did after Sarajevo, the number killed in most operations is small. During the course of the "new" terrorism the greatest loss of life has come when bombs have been placed aboard airliners. Often deaths result because of a mis-sent warning or a slow reaction time rather than by intent. For nearly a decade the events in Northern Ireland have never been far from the front page. Each year the death toll is watched as some macabre thermometer of horror. But each year the number killed on the roads in Ireland is far higher and no attention is paid. In Brazil more people are killed accidentally during Carnival Week than have been killed in twenty years by contemporary Brazilian revolutionaries. But then, the consensus that terrorism kills relatively few people is cold comfort to the victims.

Again the specialists have concluded that the "new" terror is either *a transient trend or a growth industry.* In the first case we are caught in a typical wave-pattern cycle—a rapid leap upward, an early peak, and a fading away to nothing—so that we need only wait for the fading away. Others foresee a world where a great many aspirations and desires are frustrated and where those who have no other means to act on events will resort increasingly to terror. The latter group tends to be ahistorical, and the former, sanguine conservatives; but both are projecting a surprise-free future based on their reading of recent trends.

For many social scientists these trends are not read but counted: so many incidents of hostage-taking, so many hijackings, so many no-warning bombs. There are two problems with this quantitative approach: first, the sources, even for supposedly elegant agencies like the American CIA, often leave something to be desired—some assassinations never make the *New York Times Index;* and second, it is not the number of incidents or deaths that matters but their exaggerated impact. The Croatians were responsible for one death and one incident, but they became the media-event of the week. Thus, counted incidents, even if all incidents could be counted, might well be falling off (or piling up); but it is not the numbers involved in Munich or Khartoum that matter or whether there were more or fewer incidents in 1973 than 1972. The question is not the quantitative trends but the prospect of further spectaculars—and at best the common wisdom indicates that the frustrated are unlikely to deny themselves. Of course, as a result of unforeseen factors the trends in political violence may change, or at least become less visible. No one knows.

In this connection *terrorists are likely to be more effective or else they may not.* Especially for academics the horror-scenarios have held considerable fascination. The theft of plutonium and construction of a nuclear device, wholesale poisoning of water supplies, the spreading of loathsome diseases, the use of radioactive materials—quite feasible horrors, at least in fiction or ivory tower speculations—are regularly presented as a potential reality. And even if the hypothetical terrorists eschew poison vials or black death or ten kilos of plutonium, there are ample new "conventional" devices that have caused widespread concern: miniature grenades, hand-held missiles, absolutely silent submachine guns, portable flame throwers.

Against these pessimists there are others who stress revolutionary incompetence and the futility of trading-up in weapons or horror when so much can presently be done with so little. These "optimists" are in a distinct minority. Still, in the last ten years there have only been three attempts to use hand-held missiles—all by Arab fedayeen—one at the Rome airport and two at the Paris airport. Except in irregular war situations, most terrorists have been content with pistols or submachine guns or, in the case of the Croatians, pot-bombs. So, while the terrorists *could* become more efficient and deadly, especially with the help of patrons, to date they have not felt any such urgency. But according to the common wisdom, they may at any moment.

What kind of person is the terrorist? *The terrorist is either a psychotic fanatic beyond accommodation or he is a rational rebel.* One reason for the disagreement is that few specialists have ever seen a terrorist, even at the end of a gun, and even fewer have met one who is not in prison or in retirement. Obstacles have never thwarted the academic mind, and in this instance the paucity of sources has actually encouraged speculation. At one end of the spectrum we have New York Police Commissioner Michael J. Codd, who saw the Croatian hijackers as "madmen, murderers," and Kozo Okamoto whose strange explanation of the Lod massacre lent credence in many quarters to the belief that terrorists are mad.

> The modern terrorist phenotype is a new product of our culture, existing on its own logic, for its own purposes, regardless of ideological justification. The phenomenon is a menace to society because of its new destructive capacities and a concomitant unpredictability of behavior which puts it beyond control.[2]

[2] *Survival*, July–August 1973, p. 183.

On the other end of the spectrum are those who see "terrorists" as quite rational men or women—idealists and patriots who serve in their organizations for much the same reasons that the more fortunate can volunteer for the Special Air Service or the Green Berets. They are led by equally rational men who have chosen a particular set of tactics on equally rational grounds—as the fedayeen leader said, "We would throw roses, if it would work." Unable to win in a conventional war, the terrorist's only hope is to find other means. Apologists for the various revolutionary spectaculars or no-warning bombs point out that the leaders of states have more often evinced the traits of the mad—Hitler slaughtering millions upon millions, the grotesque Idi Amin, the Greek colonels.

Much discussion focused on the mind of the terrorist goes on in isolation from the subject. In and out of analytical circles there is continued confusion of criminals, psychotics, and revolutionaries. Most hostage-bargaining strategies assume a similarity despite the obviously different priorities. If, instead of the fanatics of the Hanafi sect, Black September had been holding hostages in Washington, it seems likely that negotiations would have evolved differently. And for political terrorists there is an eagerness to fit academic models to real people: terrorism and the theory of discontent or frustration-aggression or the dynamics of individual anger and *angst*. The resulting psychological profiles tend to be academic. To contend that a terrorist is a person prepared to surrender his own life for a cause considered transcendental in value, even if true, does not reveal a great deal about an individual terrorist or the species in general. "Give Me Liberty or Give Me Death" is a political posture, not a psychological one. There is, as well, an inclination to generalize across national boundaries and cultures. When apparently accurate, at least for those revolutionaries on active service—a person who is young, most often middle-class, usually male and economically marginal—the result is hardly prepossessing.

And when intriguing—"a person for whom all events are volitional and none are determined"—there is no discernible data. The major division is between those who see the perpetrator of such violence as exhibiting behavior that is deviate, antisocial, aberrant (and thus beyond accommodation) and those who stress rationality and self-interest (and thus, openness to reasoning). As Professor Irving L. Horowitz has suggested, a terrorist is rather like a unicorn—all things to all men and real only to believers.

Then, it is agreed that while nothing can really be done about terrorism in an open society, something must. *The problem is that there is no solution.* If a thirty-million-dollar airplane can be taken over by fanatics waving kitchen pots, what good are the magnetometers, luggage searches, psychopathic profiles, and armed guards? If politicians are going to be allowed to appear in public, what protection is there against the determined gunman or even Squeaky Fromme? Some open democratic societies, like Uruguay, have been closed down by those who felt the choice was between state murder and anarchy. All threatened societies have produced advocates for special prescriptions—accommodation with legitimate demand, death to the perpetrators, an end to media coverage, ritualization of response to hijacking, armed police, new and different laws, or communication with the desperate. All of the options, however, share the assumption—generally accepted—that there is no single, no complete, and certainly no immediate solution. Yet open democratic societies cannot drift; governments must aim to combat terrorism and win. And everyone agrees at least on a single premise—that this is easier said than done.

The last assumption of the common wisdom is that *there is a common wisdom.* Surely all those dreary conferences, the long and learned papers, the elegant articles, and the assembled data run through computers have not been in vain. There *must* be more knowledge and tactical successes. Certainly, there has been a spectacular drop in hijacking. But the drop has been in the number of psychopaths and criminals hijacking planes for reasons which are only tangentially related to the problems of political hijacking. Again, the success of the Hanafi negotiations in Washington and of hostage-bargaining elsewhere indicates real progress in responding to terrorists. But the academic foundation for the technique has little to do with rigorous research—or political terrorism. Rather, it was a pragmatic discovery that the disturbed only want to vent frustration rather than harm people that led to success. The novel techniques and tactics *do* work. A quick course in hostage negotiations brings visible results—trained negotiators save lives. Yet academics still cannot define a terrorist or describe his "mind," nor can they predict even the immediate future or analyze to mutual satisfaction the immediate past. Little wonder that those responsible for the maintenance of order despair.

CHAPTER

SOCIETY AT PERIL: THE VULNERABILITIES MADE MANIFEST

At any moment, so the rumour went, some lonely lunatic in a laboratory might blow civilization to smithereens, as easily as touching off a firework.
George Orwell

What no one could deny, common wisdom or not, was that during the 1970s there *had* been a terrorist threat and that the institutions and installations of open democratic societies nearly everywhere had been endangered. As airliners were seized, diplomats murdered, bombs detonated, the public at large and particularly those in policy positions recognized that in the complex, postindustrial world a few fanatics were seeking out the vulnerable nodes, those not only unsafeguarded but also, until most recently, unrecognized. Jumbo jet airliners containing hundreds of potential hostages could be seized by men waving pots or candles and be destroyed with a few grenades. And beyond airliners, there were other potential targets such as, for example, the new generation of super-tankers, or very large container ships filled with oil or liquefied gas—vast, ill-protected floating bombs. In fact the search for vulnerabilities and the potential of more sophisticated threats became a serious academic and policy investigation. The early returns were almost always gloomy.

An Escalated Threat: High-Technology Terrorism

Whether or not terrorism was a growth industry, any future terrorists seemed almost certain to be more lethal simply as a result of trickle-down technological advances in conventional armament. No longer will assassins have to resort to mail-order rifles or urban guerrillas to homemade bombs. In the course of the last decade a variety of infantry weapons has been developed by major arms manufacturers—infrared night glasses, RPG-7 hand-held rocket launchers, M-16 and AK-47 assault rifles. As this equipment moves on-line (as it almost inevitably will), observers anticipate that these devices will come into the hands of revolutionary groups—some already have.

The German-manufactured Armburst 300 is a short-range, disposable antitank weapon. It is silent, without muzzle flash or rear blast, smokeless, without infrared signature, and capable of employment from any small room. There is also a whole class of new, small, precision guided missiles, not unlike the Armburst 300, along with the portable flame throwers, some firing a cartridge that ignites upon impact, and silent grenade launchers. The Provisional IRA has used the Soviet-made RPG-7 in Belfast, and so have the Palestinian fedayeen during attacks on airliners on the ground. Neither group did so with any great success, it so happens, for the IRA used armor-piercing projectiles improperly, and the fedayeen in Rome missed the El Al jet and hit a Yugoslavian one and in Paris were ineffectual. Various silent weapons have been manufactured—and grenades have been miniaturized and simultaneously increased in range, accuracy, and effect; a tiny, laser-guided .22-caliber submachine gun has been privately developed. Explosives have become more sophisicated and can be detonated from considerable distance, years after being implanted. Incendiary weapons like napalm exist in quantity. All these are simply *conventional* weapons, easy to convert to unconventional purpose. They may be too sophisticated for untrained terrorists; they may be unneeded by urban guerrillas, who can do nicely with homemade diabolical devices, or by assassins, who often get by with a Saturday-night special. But the existence of these new weapons does indicate an uncertain future.

The scenario builders have constructed even more dread possibilities, involving the whole spectrum of chemical weapons, nerve gas, hallucinogens, and various lethal or dangerous agents, some legally available, others that can be created by the knowledgeable. (Such scenarios are not all that improbable. In April 1975, terrorists stole fifty-three liter bottles of mustard gas from German Army bunkers and threatened to unleash gas attacks on several cities, including Bonn and Stuttgart.) Then there is the possibility of biological weapons—anthrax, typhoid, plague—and most awesome of all, radiological material like plutonium oxide or enriched uranium.

The ultimate terrorist weapon, and a major preoccupation of many, would be a nuclear device constructed with stolen plutonium or a melt-down in a nuclear facility that would guarantee widespread radioactive pollution. With the rapid growth of the nuclear power industry, not to mention the huge stockpiles of nuclear weapons (some miniaturized to the size of a cigarette pack), the vulnerabilities of the nuclear facilities have been bitterly criticized by those who fear the advent of terror on a grand scale.

It is not simply nuclear installations that appear inadequately safeguarded but all sorts of sophisticated systems. Water supplies are vulnerable—there was even a report of a neo-Fascist plot in Italy, proposing the use of radioactive substances to poison Rome. Oil refineries and pipelines have been sabotaged. A highly detailed study has been published which focuses on the prospect of industrial sabotage in the North Sea oil fields. And, of course, the huge super-tankers ply the seas with tiny crews and, in the case of the liquefied natural gas carriers, incredibly volatile cargos. Increasingly, weapons systems, communications systems, and billing and banking systems have become highly complex and also highly vulnerable.

Millions could watch but not hear one of the Ford-Carter debates because of a malfunction in two tiny bits of equipment with a combined price of fifty cents. Sabotage of computers or microwave relay towers or electric generators by the knowledgeable could cause disproportionate discomfort or danger. Just as a single infantry soldier with a handheld guided missile may threaten a multimillion-dollar main battle tank, so can a future terrorist strike with assurance of great effect by carefully selecting his target.

But it is not simply vulnerable *things*—installations facilities and systems—that more sophisticated terrorists may endanger but also the institutions of society. There are those who would destroy democracy to save it—declaring a state of siege, narrowly restricting civil liberties, greatly extending the powers of law enforcement agencies, restricting the freedoms of the press, prohibiting free speech to "traitors." Latin America's showcase democracy in Uruguay was closed down by the military because conventional means had not been able to repress the Tupamaro urban guerrillas. In Northern Ireland the British interned suspected terrorists without trial; in England legislation permits the authorities to expel suspects; in the Irish Republic new emergency legislation denies certain groups access to public radio and television. Sweden, with no native terrorists, felt compelled to enact antiterrorist legislation in 1973 designed to increase control over persons entering the country and to observe foreigners who might be planning political violence. Germany and Italy have both responded to indigenous terrorism by enacting laws that restrict traditional liberties. Nearly everyone agrees that *something* should be done, even if the remedy is some vague international convention. But there is no consensus of how to protect democratic institutions without destroying them. And no area is more vulnerable than the press and the broadcast media, often seen as manipulated to terrorist advantage and regularly defended as simply performing a historical role—reporting the news.

Vulnerability: The Media and the Masses

As the perception of the new terrorism grew firm in the early 1970s, the role of the media, especially television, came under scrutiny. The increased use of action film and the arrival of the first commercial communication satellites made possible extensive coverage of spectaculars—fires or small wars or terrorist dramas.

Those responsible in the networks in editorial offices continue to insist, certainly in public and often to their colleagues, that they, as always, simply covered the news. And assassinations and hijackings, they claimed, were certainly news. But this was not actually the case, for terrorism and media coverage existed in a symbiotic relationship. Television no longer just responded to a terrorist-event; it became an integral part of that event.

Of course, it has long been true that to a considerable extent much of the "news" is manufactured: presidential press conferences, open congressional committee meetings, campaign addresses, sports spectaculars, leads from the highly placed. And traditional notions of what constitutes a good story have not changed—personal drama, violence, suspense, and, if possible, sex. Small wonder that, as a newsworthy event, the single greatest magnet for the broadcast media and the press in the 1970s was the long-running adventures of Patty Hearst and the Symbionese Liberation Army: beautiful young girl, famous parents, weird villains, shoot-outs, the incredible transformation of girl to guerrilla, and a trial at the end, with various unsavory revelations. The coverage was massive. And why not? After all, the great shoot-out was seen *live,* in color. The SWAT machine guns were real. The blazing house might—who knew?— hold Patty. And everyone, including her parents, could watch the action.

It has in fact been the potential of instant communication that has attracted the terrorists. Through a series of trials and errors they have learned how to choreograph the ideal media event—"news" so compelling in traditional terms that the media *must* respond. For the new terrorist, then, there are not only victims but beyond them, also targets—"public opinion," their avowed enemies (traitors in the ranks, moderates, and weaklings), their own friends and faithful. In order for the fate of the victims to have the desired impact on the target, there must be not simply a means of communication but also a form that will guarantee awe, outrage, anguish, or horror. The ancient messenger bearing bad news in a forked stick, the telegram from the War Office, Edward R. Murrow speaking above the crackle of burning Wren churches during the London Blitz, these all lack the incredible immediacy of television. The tube takes you—*live*—to the Munich massacre or the SWAT shoot-out or the murder of Lee Harvey Oswald. With millions watching, two or three hostages murdered on camera become millions of hostages killed on millions of screens—truly horror on horror's head.

Over the course of television's relatively brief history, those with a cause have discovered that certain kinds of behavior within range of the camera's eye guarantee coverage: action is news. What the producers of terrorist spectaculars have managed is to raise the level of attraction to a peak beyond former riot rituals, to a level comparable at times to coverage of more conventionally global events, like the World Cup final or the Olympics.

A terrorist-spectacular first should be staged in an ideologically satisfactory locale with more than adequate technological facilities. Munich was ideal—no Justice for Palestine, no Peace for the World, not even at the Olympics, and several thousand journalists and cameramen on the spot. Even marginal sites like Khartoum, ideal for ideological reasons, can now be used as a stage, although the great Western cities are preferable. The Croatian odyssey managed to include New York, Chicago, Montreal, London, and Paris, with Reykjavík in Iceland as a throw-in. While the *entire* purpose of such an operation is not television transmission, a substantial portion is.

Second, the terrorist drama must offer the reality or prospect of violence. Unlike conventional television serials, the violence is real and the outcome uncertain. At any stage in the ritualized cycle of seizure-demand-negotiation-denouement there may be violence, and as negotiations continue, the prospects heighten—the ultimate deadline under sand clouds in Khartoum or on the tarmac at Fürstenfeldbruck outside Munich. But most terrorist-spectaculars are crafted under the eyes of the security forces. Thus, while the prospect of hostage-violence is within the terrorists' control, they have also written a part for the security forces outside their direct control—as a chorus, if there is concession, or as gunmen (actors), if there is to be confrontation. What made the Israeli commando raid on Entebbe Airport doubly dramatic was that the terrorists had not written in that role; they were as stunned as would be a theater audience if Hamlet refused to die or Macbeth won out in the last act. The classical form of the drama—the media event—had been violated.

The third component of the successful terrorist-spectacular

under optimum conditions is movement—the change of scenery that allows the cameras to follow the actors (terrorists, hostages, security people) from one site to the next—coupled with the passage of time. The Croatians ran for thirty hours, from 8:19 P.M. Friday evening until 3:00 A.M. Sunday morning, long enough to command the broadcast media for three days and the front pages from Saturday to Monday. Perhaps the longest terrorist-event was the confrontation between Irish police and the two Irish Republicans who held a Dutch businessman hostage for thirty-six days in an upstairs room of a housing estate thirty-five miles outside Dublin. In that case, and in several similar barricade-events, the stalemate was not planned. A truly crafted terrorist-event foresees confrontation and, if possible, has plans for movement to another site—and, perhaps, further confrontation. On several occasions, siezed airliners with hostages have been flown back and forth, while the hijackers sought effective sanctuary and assured extensive coverage of their exploits. Once a terrorist-event is launched before the camera, the drama by definition is a success. Operationally, all those involved may be killed (as was the case at Entebbe), or captured and imprisoned (as is regularly the case in operations inside Israel). Still, the impact exists; and in fact the impact may be greater because of violent failure, as was the case at Munich. The terrorists-actors, of course, would prefer not to die or fail. Although they are quite willing to take disproportionate risks with their lives, yet they are comforted by the excellent odds on ultimate freedom as long as they come through alive. A RAND survey of sixty-three major hostage situations between 1968 and 1974 found that seventy-nine percent of all members of the terrorist team escaped punishment or death and achieved a virtually one-hundred-percent probability of gaining major publicity whenever that was one of the terrorist goals. And it nearly always is. It does not matter if the coverage is hostile, or if the observers are outraged, indignant, and disgusted by "feral guerrillas" or "rabid creatures." Terrorists anticipate such a response from that particular target-group. At Munich Black September sought from one target—Western opinion—a single response: why would people do such a horrible thing? From their other target, their allies and friends in the Middle East, they sought and largely received understanding. Why should the Arabs care about Western anguish? "They never cared about us. Why should we care about them?"

The terms of the success have been difficult to measure. Certainly, in conventional terms the "issue" has been raised—Palestine, South Molucca, Croatia, if not exactly household words, now have a substantially higher recognition factor. Secondly, the concept of no-peace-for-you-without-justice-for-us has brought home to many of the comfortable seeking an easy life that the world of the immediate future may be an uneasy place. Justice for all appears impossible. And if justice is impossible, so too is peace for all.

Finally, the various revolutionary organizations have found a purpose; their members and supporters have been encouraged that even if their aspirations are no closer to reality, they can at least still act on events. Certainly, gunmen, hijackers, and bombers must feel such operations effective, for despite the public indignation and outrage, they continue on their allotted way, shooting enemies of the cause, stealing airliners filled with innocents, setting no-warning explosive devices in bars and department stores.

And so far the media has been quite willing, usually eager, to broadcast those events; editorial writers and reporters dwell on them at length. Only rarely is there dissent that this is news fit to print:

> We of the Western Press have yet to come to terms with international terror. If we thought about it more and understood its essence, we would probably stop writing about it, or we would write about it with a great deal more restraint.[1]

[1] *Washington Post*, 21, November 1975.

About the only adjustment to traditional coverage—everything that space will allow for a hot story—has been tentative suggestions that a bad press or nasty coverage will dissuade potential terrorists. Such a "reform" is based on a failure to understand that the quality of the coverage is quite immaterial to the terrorist's purpose; only the intensity and quantity of coverage matter—the minutes of prime time, the size of the headlines. In any case the news industry does not *want* to limit coverage, even if continued coverage might encourage governmental intervention—that is, censorship of sorts. A. M. Rosenthal of the *New York Times* speaks for many when he insists that it is not the responsibility of the media to accept such self-restraint. About the only publicly expressed unease with the coverage of terrorist spectaculars is the degree of cooperation with the police. There are nowhere hard and fast rules. At times the British have used D-notices to prevent coverage related to national security, but such a convention can work only in a small country with an interchangeable elite; it would be more difficult in countries with not only a free press but also one founded on profit and/or beyond much governmental control. So the involved have tended to narrow the scope of their self-examination to the traditional questions of coverage response rather than grasping the nettle. In sum, the media have not "responded" to the terrorist threat but instead have become an integral part of that threat.

The arguments that terrorist-spectaculars must be covered—live if need be, at great length if warranted—have been traditional. The public must be informed—to hush up an incident would erode the trust of the American people in their democratic institutions ("What are the news people keeping from me?"). In any case potential television terrorists frustrated by the imposition of any such ban might well devise a more awesome media-event that would force coverage—an escalation of horror. What the terrorists have discovered, of course, is that television news does more than inform. The "news" in the West has always stressed sensationalism and novelty. The media entertain.

There is simply no way that the Western media can ignore an event that has been fashioned specifically for their needs. Television terrorists can no more do without the media than the media can resist the terror-event. The two are in a symbiotic relationship, so that any restriction of one narrows the bounds of the other. To be free, the media have to be captured. And the media have been captured, have proven totally defenseless, absolutely vulnerable. Of all the foundations of a free democratic society, that most basic—the freedom to know, to be informed—has guaranteed that such knowledge and such information can be fashioned by the fanatic through the conduit of the media eye. To close that eye would erode a fundamental right, would close an open society. Yet not to do so assures future massacres, further terrorist-events with little hope of audience saturation—after all, people still go to see *Hamlet,* and there they know the ending.

Vulnerability: The Nuclear Option

While the vulnerabilities of the media have been exposed by the television terrorists, there are other threats that remain largely speculative. The scenario builders, secure in ivory towers or government offices, have stressed the special danger inherent in a high-technology society. For the specialists the vulnerabilities are so obvious, so dangerous, that very real safeguards are needed, even if the threat remains theoretical. The nuclear option was the greatest single concern of many long before terror became trendy at the very beginning of the atomic era. On Arpil 25, 1945, Henry Stimson sent a memorandum to President Truman on the matter:

> . . . the future may see a time when such a weapon may be constructed in secret and used suddenly and effectively with devastating power by a wilful nation or group of much greater size and material power. With its aid a very powerful unsuspecting nation might be conquered within a few days . . .[2]

Some few felt that any such "group" could be very small and that the vulnerabilities of the industry were far greater than most assumed. Their number grew.

Out of public view various concerned government agencies commissioned studies, mostly classified, that concluded existing safeguards were inadequate to prevent theft of fissionable material by those whose resources were no different from past thieves intent on stealing other valuable materials. It was felt there was a growing likelihood that nuclear material might be stolen by non-governmental organizations. As yet, however, the debate was either classified or, in academic circles, confidential, for fear of assuring a self-fulfilling prophecy.

Essentially, there are two vulnerabilities: theft and sabotage. In the latter case the increased number of ill-protected nuclear industry sites, military bases, and AEC facilities of various sorts might prove tempting targets for symbolic sabotage or attack by the demented. Perhaps no one wanted a core meltdown and the ensuing radioactive fallout that might endanger hundreds of thousands of lives, but if a major sabotage incident was not probable, it *was* possible. In the case of theft, either secretly, by someone within the nuclear facility, or as the result of an armed raid of terrorists, the prospect was a radioactive dispersal related to blackmail or simple destruction, and/or the construction of a low-yield nuclear device.

At this relatively early stage much of the discussion took place either in classified circles or within the nuclear establishment as confidential exchanges, often in arcane lan-

[2] Morton Grodzins and Eugene Rabinowitch, *The Atomic Age* (New York: Basic Books, 1954), p. 7.

guage. Although safeguards had become a considerable problem, it was a professional problem, more linked to engineering than politics. The same concepts, often poorly defined, appeared on various agenda, with the focus on the matter of material balance (MUF—materials unaccounted for—lost, strayed, or stolen?) but with little progress in analysis or in policy recommendation, although extensive scenarios were constructed. While there was not a sense of complacency, there was, some felt, a lack of urgency.

Theodore B. Taylor, for example, argued that the acquisition of the appropriate nuclear material and the construction of a nuclear device was all too possible, not beyond the capacities of most potential terrorists. Taylor estimated that nuclear terrorists would need, for example, no more than four kilograms of plutonium or eleven of high-enriched uranium; and with skill and appropriate equipment they could well make do with less. It was even possible to purchase all the necessary theoretical and reference material from the Government Printing Office or traditional scientific sources. According to Taylor, even a lone individual with an average undergraduate science education should be able to design and manufacture a crude fission bomb that would on detonation produce a yield equivalent to at least one hundred tons of high explosive and also would be small enough to be transported in an automobile.

To test Taylor's theory, project "Nova" at MIT assigned a student the task of designing a workable nuclear device using only available library references. According to Dr. Jan Prawitz of the Swedish Ministry of Defense, a bomb constructed from the resulting design might or might not go off. And in October 1976, a Princeton student, John Aristotle Phillips, wrote a report demonstrating how an atomic bomb can be built for less than two thousand dollars. In May 1977, the London *Express* claimed that an employee had constructed a nuclear device, except for the necessary plutonium. While there remained doubts about the designs, it was generally assumed that if terrorists had more advanced training, the prospects of a device with the yield of approximately one-hundredth of the Hiroshima bomb would be realistic.

Taylor's colleagues, however, continued to view the prospect as rather alarmist, so he decided, in effect, to go public. With Professor Mason Willrich, he prepared to publish under the auspices of the Ford Foundation. Simultaneously, in 1973 and 1974, the General Accounting Office published reports critical of AEC requirements for physical security of nuclear material and claiming that there was inadequate protection during the transportation of nuclear materials. In Washington in 1974, Senator Abraham Ribicoff released a consultant's analysis—"The Rosenbaum Report"—urging that attention be given to the real problem of terrorism. Later in the year Taylor and Willrich's *Nuclear Theft: Risks and Safeguards, A Report to the Energy Policy Project of the Ford Foundation* was published—including a paperback edition.

Increasingly, a far broader public became concerned, especially because the future of nuclear power had become a serious political issue. By 1975, the "terrorist threat" had become a major argument of those opposed to nuclear power. When Taylor first became concerned with the problems, low-growth idealists targeted on nuclear power as technology out of control, arguing that nuclear power stations were (1) a danger in peacetime—there was a horror list of near disasters and endless accidents; (2) an incredibly vulnerable target in wartime, especially when sited in urban areas; and (3) a continuing temptation to terrorists, to criminals, and even to malignant individuals. The advocates of nuclear power were outraged; stealing plutonium and building a bomb was not *that* easy, security was improving, all sorts of safeguards made such an eventuality improbable. Scenario-building that centered on what *could* happen, however, was more attractive to the speculative, academic mind than was the task of explaining why things would not happen. David Krieger, in an article entitled "What Happens If . . . ," (published in *Annals* March, 1977), offers several short scenarios of potential disasters:

> A U.S. army base is destroyed without warning by a low-yield nuclear weapon with no clues as to who is responsible.
>
> A cadre of revolutionaries, including a nuclear engineer, take over a nuclear power plant and threaten to initiate a core meltdown if their demands for policy change are not met.
>
> Japanese extremists dive bomb an American nuclear reactor, causing a meltdown.

The advocates of nuclear power could only point out that the scenarios *were* academic and that the threats and violence were *very* low-level—and in some cases apparently motivated by the ideals of the environmentalists.

Basically, both ends of the spectrum agreed that theft was possible, that a nuclear device could be constructed, and that both sabotage and/or radiological terrorism could occur. Even the responsible agreed that a terrorist attack was possible. Manning Muntzing, former Director of the AEC, noted:

> A band of highly trained, sophisticated terrorists could conceivably take over a nuclear powerplant near a major city and destroy it in such a way as to kill thousands—perhaps millions—of people.[3]

Terrorists might bomb facilities from the air, raid them in commando teams and barricade themselves inside, stage bazooka attacks, or drop explosives from helicopters—and rarely would there be an effective defense. There had been the regular attacks on nuclear installations or facilities. In the United States there had been the threats to nuclear facilities; and in Europe various radical, covert groups—German, Breton, Swedish—had undertaken symbolic sabotage. A German member of parliament had even carried a *Panzerfaust* bazooka into the Biblis reactor and presented it to the

[3] *Los Angeles Times*, 16 October, 1974.

director. There must, therefore, be adequate safeguards; but safeguards that would satisfy the critics of nuclear power were beyond existing resources. Their conclusion was that the nuclear power industry should be phased out and that fast breeder reactors, which would produce more plutonium than they used, should not be built at all. In the latter case the new Carter Administration agreed and budgeted no further funds for the Clinch River breeder reactor. From a faint possibility, raised by very few within the establishment, the terrorist threat had become a determining factor on the priorities of American energy options.

The defenders of nuclear power felt that they were being defeated with a spurious argument. In a vague and contentious field there has been no evidence that a terrorist threat exists. Second, all the "terrorist" attacks on nuclear facilities have been symbolic or are so minor as to be irrelevant to the argument. Third, and most important, there is no evidence that terrorists have any interest in killing large numbers of people with a meltdown. The new transnational television terrorists want media exposure, not exposure of the masses to radioactive fallout. And finally, the technological capacities of organizations with sufficient military skills to launch an attack—for instance, the IRA—are not great. The mix of motive, military and technological skills, resources, and perceived vulnerability simply does not exist.

The case for individuals or the disturbed or criminals has even greater flaws. Mad bombers do exist and have been attracted to nuclear sites; but one man alone would have to devise a means to divert the plutonium, to have considerable technological skill, to possess adequate resources, and to be determined on constructing a nuclear, rather than conventional, device. The advocates of radioactive-malevolence stress the irrationality of the threat and ignore the limitations such irrationality places on a reasoned progress toward a one-man nuclear capacity. The criminal connection assumes a degree of sophistication and investment in a novel area that seems unlikely. Criminals, organized criminals, tend to be conservative, both in their professional activities and their life styles. To enter a novel field abounding in vast challenges and almost certain to trigger a violent international response by the threatened would be daunting. The diversion and sale of small amounts of radioactive material might appeal to individuals or even organizations if the risks were small, detection unlikely, and profits high. But the idea of a market suddenly opening up for a strategically significant amount of nuclear material for Corsican or Mafia salesmen to exploit remains improbable.

In sum those who might have the desire—the crazy bombers—tend to lack the capacity, and those with the capacity lack the desire to kill the many. Yet under certain undefined circumstances the hypothetical threat proposed by the scenario builders might become a reality. At that point the degree of vulnerability and the nature of the safeguards would prove crucial.

Those in the private sector of the American nuclear industry were appalled at the rising costs of security and the necessity to counteract the most elaborate terrorist scenarios. They were deeply concerned when the Carter administration decided against funding the Clinch River breeder reactor plant and insisted that new and improved security measures reduce the prospect of sabotage or theft, and that further technological safeguards could be developed. For example, for four years in the 1960s a breeder nuclear-fuel recycling system in which the fuel remained so radioactive that the terrorists could not have stolen it operated in Idaho. It might be possible in time to keep the entire fuel chain, even in transport, in a form impossible to steal. A variety of other security technologies had also been brought on line. In May 1977, the French announced a safe way of producing nuclear reactor fuel that can be used to run power stations but not to make nuclear explosives. Certain problems, of course, might remain—might always remain.

Thus in a curious way the vulnerability for the most contentious of all Western installations—the nuclear target—has engendered not only the most detailed examination but also the most extensive investment in safeguards. Nuclear facilities, except for the crown jewels or the gold in Fort Knox, are perhaps the *most* protected, more so than microwave towers or hydroelectric dams or even political leaders. Yet atomic *angst* persists; for despite efforts to limit the plutonium economy, to control the fuel cycle, to increase safeguards, to deter any potential threat, and to fashion a world self-denying consensus, the ill-defined threat of terror on a grand scale will remain. Just as, by their very nature, the media in a free society are vulnerable to exploitation, so too is any high-technology, open society. The threat of nuclear terrorism may be problematical, but the vulnerability is real, the more so because of public efforts to assure adequate safeguards. If nothing more, the advent of the new terrorism revealed all too clearly that not only does the West live in the interesting times of the Chinese curse but also in perilous ones that offer no absolute security.

CHAPTER

SOCIETY AT PERIL: THE TECHNIQUES AND TACTICS OF SAFEGUARDS

Thou shalt not be afraid for any terror by night; nor for the arrow that flieth by day.

Psalm 91

For the vulnerable West it has often seemed that (even excluding the prospect of a nuclear threat) Providence might be the only protection. Moshe Dayan has suggested that terrorist incidents more closely resemble natural disasters than acts of war. Even then, the wise man, trusting not solely to Providence, can take certain precautions if he lives in a cyclone zone: build effective storm cellars, invest taxes in meteorological prediction services, and prepare contingency disaster plans. Of course, those who live along earthquake faults or in the predictable path of hurricanes might simply move; but they don't.

Thus those responsible for the threatened have had to find appropriate means of response to the unnatural terrorist calamities. Except when escalation created a real insurgency, as became the case in Northern Ireland, the old and detailed response to the guerrilla challenge by democratic governments seemed to be of little use. What use were counterinsurgency tactics in preventing the Croatian hijacking, in countering the rape of the media, in deterring the potential nuclear terrorists? At least in Northern Ireland, the British army and Ulster constabulary have played almost traditional roles; they could deploy in traditional ways, although employing novel equipment and more refined tactics. The terrorist challenge, especially the sudden spectacular drama of the transnational television terrorist, required a novel mix, with the military as only a single (and often least significant) component.

At the very first, in Lod, Munich, and Khartoum, the question of immediate and ultimate responsibility arose. Was Israeli airport security adequate at Lod? Could there have been forewarning? And what happened to the security on boarding Air France Flight 132 at Paris and Rome? At Munich there were a variety of German police forces involved, as well as local and national officials. The Israelis were present as unannounced advisors, and various foreign governments were contacted. The tactical disagreements on the ground, especially between the Israelis and Germans, were considerable, reflecting quite different priorities and options of response. At Khartoum, other than the Sudanese, Saudis, and Belgians, the "Americans" had to carry on negotiations at great distance and the ultimate step was a televised presidential press conference rather than more immediate contact. Because a terrorist incident may be a police matter or call for expert psychological advice or take place at a distance and with muddled authority on the ground, the very first problem became control at the center.

The spectrum of control devised by the threatened states varied considerably. In Britain during various IRA bombing operations and hostage negotiations, the Home Minister insisted that, as always in such criminal matters, the chief of the metropolitan police—the head of Scotland Yard—had the ultimate responsibility. And in considerable part he in turn left the major responsibility to the appropriate division. During the Munich spectacular the heads of state of Germany, Israel, and (to a degree) Egypt became involved and were the ultimate authorities. In the case of Khartoum, Nixon refused to negotiate, hewing to a policy line that eliminated the need for flexibility. In the Croatian case the French government decided on the hard line for reasons quite different from President Nixon's, but with as little contemplation. In general, even when a special terrorist control center was set up, it was often by-passed by spectacular events and control ended up in the Oval Office or the presidential palace.

By no means does this mean that the efforts to respond to the terrorist challenge by creating new forms were simply a political attempt to fashion something between the police and the president. Even in Britain, where terrorism was considered solely a matter for conventional law enforcement, Scotland Yard had a special antiterrorism organization within the force. The Germans and, increasingly, others too created not simply another squad specially charged with a terrorist responsibility but new forms employing new tactics and techniques. The British might feel it rarely necessary to arm their police but they had never had a Munich shoot-out. Almost inevitably the violent incidents revealed a lacuna in the forms of response. During the challenge of *Le Front de Liberation du Quebec* (FLQ) in Canada, the central government in Ottawa discovered that the provincial government was not going to be able to cope and that seemingly the only alternatives were the Royal Canadian Mounted Police, alone hardly likely to be effective, or the army. The federal government invoked the War Measures Act and sent in both the Mounted Police and approximately twenty-thousand troops. The Canadian experience was not unique. Consequently, the new forms tended first to institute centralized control—like the United States Cabinet Committee to combat terrorism; second, novel law enforcement responsibilities concentrated in a single division—the Italian antiterrorist unit; and third, and more visible, new security forces with new talents and new responsibilities—the Israeli Wrath of God.

If there was general disagreement on the appropriate bureaucratic response to terrorism, there was less disagreement on techniques, technologies, and tactics. Those who felt the need of a sniper could easily agree on the nature of his weapon and the rigors of his training—it was almost as important that a sniper fire *only* on specific and reasoned orders as it was for him to hit his target. Some of the tactics were major factors in an antiterrorist strategy: the Israeli retaliation policy incorporated into the *Mivtzan Elohim* (Wrath of God); some were significant in responding to special incidents—the evolution of hostage bargaining techniques; and many were the applications of useful technologies.

Response: Tactics, Techniques, and Technologies

Crisis Management

Quite often the threatened had no structured bureaucratic means to respond to a terrorist incident until after it was too late. And once such an organization was established with appropriate authority, its control during the incident was often uncertain or its existence simply ignored by those with real power. Thus in the United States the Cabinet committee is a bureaucratic response to terrorism; but, for example, in New York City unless a federal statute has been violated, thereby involving the FBI, the local police are responsible for law enforcement. Even in a federal matter, like the Croatian incident, the Federal Aviation Administration and/or the State Department may be responsible rather than the FBI. Plans for emergency readiness are even more chaotic; for example, in 1977 in the United States, 175 interagency committees and groups would be involved in the case of the nuclear terrorist incident. Sometimes, as was the case with the Croatians, there was little that "central control" in the FAA building could do but monitor events—and broaden a bit the rigid no-negotiations-no-concession posture of the previous administration. Basically, any spectacular terrorist incident is so imbued with highly volatile political issues that control is often seized by the top and decisions made for reasons of state policy rather than as "effective" tactics to achieve a "satisfactory" outcome. The result nearly everywhere is that the new bureaucratic forms of crisis-management have not worked particularly well. In some cases, when there is a consensus, they have been able to educate those who fashion political policy, even if they are later ignored at the moment of decision. In others the special antiterrorism office or the coordinating committee has been, as probably intended, simply window dressing.

New Forces

Far more important in many cases have been new forces that are concerned not with command, control, and coordination but with tactical tasks—federal air marshals, special commando groups, hostage bargaining teams, or bomb squads. At the most local level, law enforcement agents on their own or at the suggestion of the national government established special groups—seconding police into a new squad to deal specifically with some aspect of the new terrorism.

In New York, for example, in 1972 Police Commissioner Patrick Murphy decided that the almost traditional policy of "shoot first and ask questions later" had to be revised. After Munich, Deputy Chief Inspector Simon Eisdorfer, head of Special Operations Division, assigned Patrolman Harvey Schlossberg, who had earned a Ph.D. in psychology largely at night school, to set up a program on hostage situations at Floyd Bennet Airport for several hundred officers. Largely avoiding orthodox psychological labeling and making use of common sense and a few basic principles, Schlossberg put together a course manual and exposed the officers to the concept of delay, negotiation, and patience instead of recourse to the gun. The result was the Detective Bureau Hostage Negotiation Team, consisting of seventy members with linguistic and psychological resources, charged with negotiating in hostage situations. In New York's case, the incidents involved criminals or psychopaths; but elsewhere similar teams, often using the Department's own tactical manual, were concerned about potential political terrorists.

At the same time the police nearly everywhere expanded their capacity to respond to armed and determined terrorists, often intent on confrontation. The Special Weapons and Tactics—SWAT—groups found a much more congenial reception with most law enforcement agencies—the romance with automatic weapons, the distinctive uniforms, the paramilitary tactics, even on occasion the use of armor. The two prime arguments against special tactics units of the police, especially in the United States, has been that such a hard response is not always effective and is seldom needed. New York City's armored personnel carrier, acquired in 1968, has been used only twice. The first time was on January 19 and 20, 1973, when four Asanti Black Moslems held eleven hostages at John and Al's Sporting Goods Store in Brooklyn (this was also the first time that Schlossberg's hostage negotiations tactics were employed). And the second incident came outside the city, in Westchester County in 1977, when Frederick Cowan, a Nazi sympathizer, was besieged for several hours. The police in Westchester, nevertheless, decided to acquire a modified Brink's armored truck—just in case. Most important, the militarization of the police produces unnecessary, even counterproductive, "soldiers." In Los Angeles SWAT members have been reduced to providing security at the Academy Awards while clad in tuxedos. The idea has, nevertheless, caught on, and by 1977 there were over three thousand SWAT units in the United States. Some have a low profile (for example, the Special Responses Units in Alemeda County, California, or the Emergency Service Unit in New York City), while others are togged out with black suits, carry submachine guns, and insist that the SWAT mystique discourages criminals: "We want them to think we're mean."

Even when law enforcement personnel are not transferred into a special unit, there have been established—again, first in the United States—a whole range of courses, seminars, schools, and conferences focused on the challenge of the terrorist. Some organizations have been a bit tardy. For example, not until April 1976 did the American FBI set up at their training headquarters in Quantico, Virginia, a proper group (TRAMS—Terrorist Research and Management) under a grant from the Law Enforcement Assistance Agency of the Justice Department. Increasingly, however, law enforcement officers met in international conferences, exchanged experiences, and borrowed lecture notes in an attempt to inform their personnel of agreed responses. Special tactical units might be a frill in most American cities, but most national police wanted such an option. Many, of course, had riot police with much experience in traditional confrontations; but riot training, like antiinsurgency training for the army, was of little use during most terrorist incidents.

Within the military there was a growing concern that, as an institution, the army should have an agreed response to the terrorist challenge, with accepted tactics and forms. Many democratic armies had already fought irregular wars and were familiar with low-intensity conflict in Saigon or Algiers or Aden. The British in Northern Ireland could adapt some of their experience, but elsewhere matters were not so simple. At times the Israelis simply selected the team for special missions from within the establishment, as was the case with the helicopter commando raid on Beirut Airport in 1969 and the commando raid at Entebbe in 1976. The United States, instead, gave the military a new mission that required new training programs for specific units, so that, if a rescue mission were necessary, the force would be ready. Others, like the British, transformed an elite commando unit—such as the Special Air Services—by the introduction of a new mission into a special antiterrorist force. Thus in Northern Ireland the SAS have been discovered in civilian clothes, armed with the Armalite, the IRA weapon, setting killer-ambushes. And the most novel and most criticized response to terrorism has been the creation by the Israelis of the Wrath of God state terrorists.

Although there are some within the various military establishments who feel that the army should have a role in such activity, most observers want the response to remain with the conventional law enforcement agencies as long as possible. Sometimes, of course, this is not possible, as was the case in Northern Ireland in 1969 when a police state ran out of police; still even there the introduction of the British Army as a peace-keeping force ultimately proved counterproductive.

It is clear that as the number of discrete incidents increases, so does the pressure for a military commitment. In Venezuela it was Bentancourt's refusal to turn to the army when the FALN were killing five hundred police in five hundred days that salvaged not only his government but also democracy. The pressure remains, however, especially in those countries where the military has no other exciting role, no colonial war, no prospect of an expedition. Training in peace-keeping, in riot control, and later in the tactics to counter the urban guerrilla has been introduced. Some commanders, like the police who want SWAT teams, think such a capacity is nice to have and may be of comfort in troubled times. Thus a great distance remains between Frank Kitson, who in *Low-Intensity Operations* argues for a vital military role, and Sir Robert Mark of Scotland Yard, who wants nothing more than a conventional law enforcement response to crime.

If new forms of response to terror have been created (and criticized) in most democratic countries, the same is not true internationally. Regionally, there were some institutions that could be expanded to consider any terrorist threat: safeguards have been considered in Euratom, and NATO has held conferences and contemplated vulnerabilities. The Western European countries have largely agreed to legislation that would hamper terrorist flight and sanctuary. The law enforcement profession meets more often, usually revealing the great variance in their problems and the impossibility of achieving a united response founded on a common wisdom.

Since there is a communality of concern, the lay public may assume a broad international effort. But most of these steps have been tentative—whether legislative or bilateral. The old forms like Interpol are, again by the general public, overrated. Interpol is simply a clearinghouse for information, a wire service, specifically prohibited from involvement in political crime. Where international cooperation has been effective, not form but mutual interest has been present: information on suspects passed along, intelligence leaked to the threatened, meetings between those who seek to stop arms from illicitly leaving one country and those who seek to stop them from arriving in another. Within various training establishments, those with experience lecture to those who seek to avoid similar experience. So far, however, an Anti-Terrorists International does not exist and, given the varying priorities and clashing aspirations of even the most friendly states, such an organization seems unlikely to become a reality.

Private Forces

As the targets of terrorists broadened, it became increasingly clear that the priorities of governments, individuals, and private institutions often differed, sometimes dramatically. In some cases the government might direct the nature of an industry's response to terrorism. Thus the Nuclear Regulatory Commission would not grant licenses unless adequate safeguards were undertaken by private industry: more physical security, more personnel screening, more armored guards. This was almost standard operating procedure. But other and novel problems soon arose. Governments banned not only negotiating with terrorists but letting anyone else do so. In Italy paying ransom is against the law. After their local Venezuelan director had been kidnapped and Corning Glass accepted demands that it pay for the publication of a revolutionary manifesto in elite newspapers, the government seized the company. In Tanzania, when the American ambassador, W. Beverly Carter, aided the parents of university students kidnapped by revolutionaries to pay the ransom and secure their release, he was denied an expected promotion by Secretary of State Kissinger for violating the no-negotiation restriction. This American hard line softened once law-and-order Nixon resigned the presidency, even to the extent that President Carter made the promised telephone call to the demented young man who wanted all the white people to leave the globe within a week. Thus by 1977 in America, increasingly there were both negotiations and concessions, under certain conditions.

Still there is no doubt that government and private interest are not identical. The threatened, whether tourists or transnational corporations, had to reevaluate their attitude toward the new terrorist threat. Some corporations decided to pay ransoms in those countries, like Argentina, where kidnapping executives had become a revolutionary cottage industry that produced vast sums of money. Others announced as a set policy that the corporation would *never* pay, the executive victim being left to his own resources. Some tourists avoided combat zones, others wandered in assuming the disorder had been exaggerated. Everyone had to make choices, examine relevant safeguards, and at times bring in the experts.

The result has been that the private security business has

boomed. Bodyguards are in demand. Italy has an important kidnapping at least once a week; the rich, and many politicians and prosecutors, never move without the heavy man in the dark suit. In America it is possible to rent a film on "Kidnap—Executive Style" (twenty-five minutes, ninety dollars a week fee); to take a course in protecting top officials while driving ("Sixty percent of all terrorist attacks take place around an automobile"); and to attend seminars on the terrorist threat to businessmen and the need for deterrents. Physical security has been increased—although not to the level required in government buildings—so that no longer can one easily wander into corporate headquarters. Vulnerable facilities—refiners, tankers, corporate jets—have been protected against the terrorist threat. Some companies no longer announce the arrival of their chairman in Beirut or Buenos Aires, no longer fly jets with a big, bright company logo, no longer even carry luggage with company tags. All this at the request of an expanded security division. It is clear that what may be good for the big transnationals and the errant tourists may not be good for their government—consequently, like terrorism, the new forms and priorities of private security appear to be a growth industry.

Intelligence

Nearly all of the threatened or their experts agree that the key to an effective response to terrorism is good intelligence and that such intelligence is difficult to acquire. There are in every democratic society the limitations of law and restrictions on the nature and type of penetration possible. There are as well the difficulties of operating against transnational terrorists, against those based abroad, and against those who belong to tiny, almost unknown, organizations. Tactical intelligence, indicating who is going to do what and when, is very elusive. Even the strategic intelligence focused on general trends and potential threats presents severe problems. A few countries have set up special units; and all have expanded the role, both domestic and foreign, of intelligence-gathering services. There have been, as well, new means, new methods, and new technologies employed within the new forms.

Increasingly at most airports, the filtering process makes use of dossiers on the known members of active revolutionary organizations. When Abu Daoud, the suspected leader of the Munich massacre, flew into Paris as Youssif Hanna Raji to attend the funeral of a PLO leader—shot, it is assumed, by the Wrath of God—the Israelis, among others, tipped off French security officials. Several countries also maintain in the major airports agents with total photographic recall, constantly searching through the crowds for the faces they have memorized from old photographs, newspaper clippings, and identikit mock-ups. One of the more notable triumphs of the new intelligence was the odyssey of the IRA arms ship "Claudia." The British, by a variety of conventional means—and with the traditional help of informers and agents in the Middle East—along with certain technological novelties, traced the wanderings of the vessel to Irish waters, where Dublin's navy seized the ship. In fact British intelligence coups concerning Irish arms deals in Europe have forced the IRA Council to depend largely on weapons smuggled out of the United States in odd lots—and that in turn has engendered close cooperation between American, Irish, and British law enforcement agencies. In intelligence, just as in the creating of international forms to combat a common problem, the difficulties remain constant. Bilateral and regional cooperation are often possible; special relationships exist between certain police and intelligence forces, often apparently more intimate than with their own government. No single international intelligence organization exists, or is likely to exist. Spectacular coups are rare, and even the surveillance of suspects is sporadic. Two of the five Croatian hijackers had previous records, but the FBI simply did not have the resources or the legal right to maintain surveillance—no matter what the Yugoslavs thought. Stretched thin, dealing with tiny, often unknown groups, the responsible are charged with accomplishing tasks beyond their present capacity to do so.

New Technologies

If the new forms of control or intelligence have not really solved the old problems, an area of great innovation and effect has been the development of mechanical devices that make the mission of any terrorist or urban guerrilla—or mad bomber—far more difficult. Just as there has been official concern about high-technology terrorism, there has been developed a high-technology antiterrorism, in some cases adapted from civilian or military advances, in others developed solely for the task at hand. In a proving ground like Belfast in Northern Ireland, the contrast between the IRA (with their semiautomatic Armalite assault rifles, and their homemade and unstable bombs put together out of fertilizer and stolen gelignite) and the sophisticated armament of the security forces is so asymmetrical that the British advantage is almost cancelled out.

Often the most visible technology has been riot-control equipment: gas masks, rubber bullets, CS gas, plastic masks, armored vehicles that spray identifying dye, and videotape for later identification. The average citizen is aware of the magnetic X-ray sensors at the airports, if not of the watching intelligence eyes or the use of psychological profiles of hijackers. But all sorts of equipment has been brought to play on the problem. In Northern Ireland the British have used specially equipped helicopters—*Nightsun,* with large, superbright searchlights; and *Nightshout,* with loudspeakers. Other helicopters have been equipped with telescopes by day and light-concentrating viewers at night. A helicopter with amplifying equipment that can broadcast sound that disorients listeners has been developed, and work has been undertaken on equipment to produce flicker light patterns that would have the same disorienting effect. In fact British experimentation and employment of techniques of sensory deprivation and disorientation during interrogation led not only to new skills but also to a formal charge of torture before the European Court of Human Rights.

Many of the antiterrorist techniques are simple adaptations—Ultra High Frequency communication, computer retrieval, radio monitoring, voice prints, miniaturized means of communication, miniature weapons. Some of the devices might give James Bond pause. Palestinian fedayeen leaders in Europe have answered their preprimed telephones and, at the other end of the line an Israeli, thus assured that the target is available, sends a radio signal through the wire that detonates the headpiece and the fedayee's head. IRA men have been shot at great distances in the dead of night by soldiers equipped with the new light-concentrating telescopic gunsights—it is very difficult to accept that others can see in the dark. Bombs have been discovered by gelignite sniffing machines—and by dogs.[1] Bombs have been detonated in bomb factories by remote control signals or in place by robot machines. And all those elegant weapons that the concerned feel may fall into the wrong hands are for the moment employed by governments.

Yet with sixteen thousand troops on the ground and the police and the militia with all those elegant devices, the IRA can still devastate the center of Belfast. The Prime Minister of Spain was killed when a homemade bomb detonated under his limousine, as was the British Ambassador in Dublin. Judges and prosecutors, and their body guards, have been shot down with conventional small arms in Italy and Germany. Despite everything, Hanafi fanatics waving bolo knives can seize over a hundred hostages in the middle of Washington, five people with kitchen pots can commandeer

[1] As an aside, in case of a nuclear fallout incident, rats are the only animals that can smell radiation.

an airplane and involve the governments of the United States, Canada, Iceland, Britain and France. It is no wonder that the outraged and indignant public feels that little has been accomplished.

Yet the response both to hijacking and to hostage situations as much as any other area has seen impressive progress. All the new forms and forces and technologies—and psychological analysis—have been brought to bear with demonstrable results.

The early questions of those attracted by the problem of hijacking focused on (1) what was happening, (2) who was involved and why, and (3) what could be done. In general only the immediately threatened—for instance, the United States with the Cuban connection—were concerned, until hijacking became a worldwide trend with every airline—even the Russian Aeroflot—a potential target. Usually each nation tended to view the hijackers as representing a special kind of threat. The Israelis, for all purposes involved in a twenty-five-year war with the Arabs, saw the Palestinian seizure of their planes as an act of war tolerated or encouraged by other Arab states. Their response, consequently, was military and not limited to reprisals against the elusive fedayeen. In the United States hijacking was air piracy, a crime. There was no military, and rarely any revolutionary, content. Gradually American experts came to accept that there were two main groups involved, criminals and psychotics, who often used revolutionary rhetoric. Other nations, when their airliners were stolen, felt that the act, probably political, had no real relevance to Germany or to Britain or to France. Their major preoccupation was to get back the plane and passengers and hope that the terrorists would go elsewhere. In each case, although views changed, the particular response was determined by the national perception.

An obvious first step for everyone was an increase in airport *physical security.* It was possible and remained possible in many international airports to drive a vehicle across the tarmac to the airliner—not to mention attacking it from a distance on takeoff. There were few armed guards, few guards at all. The only inspection was at customs and that increasingly rare or optional for passengers who could choose a green door, claiming nothing to declare. There were no special forces, no prior disaster planning, and no expectation of trouble. Nearly everywhere and certainly in most democratic countries security was greatly tightened as the roll of incidents grew longer. Guards were moved in. Planes were made more inaccessible from the airport buildings. Some airports, like Belfast or Lod, were very tightly secured. Others, like Rome or Athens, remained porous, in part because of the nature of their construction.

A much more important step was filtering the potential passengers through inspections—if at all possible, at considerable distance from the boarding gates or airplanes. X-ray and magnetic sensing devices were introduced. Hand luggage was examined. During the last quarter of 1972 in the United States, 1,906 passengers were prohibited from boarding. There were 1,181 arrests and a total of 3,396 weapons were confiscated. By the first quarter of 1973, the totals were down to 617 denied boarding, 573 arrested, and the weapons seized up to 4,916. A criminal now had to be quite cunning and clever to slip by with a weapon.

Elsewhere in the world filtering was even more stringent; for example, all luggage went through inspection and then was not put aboard the plane until it had been claimed from the tarmac. Thus, a suitcase bomb could not be checked through unless the passenger went with it. On flights into Belfast, no one could carry as hand luggage even a hardback book—outraged, Enoch Powell ripped the cover off his flight reading. The painstaking inspections and preparations of El Al Airlines became legendary: typewriters were often dismantled (and returned in better condition), radio or tape recorders or shavers tried out or taken apart, pills or hairsprays or alarm clocks gone over most thoroughly, books opened, toothpaste tubes squeezed. Psychological profile experts were standing by—not to mention sturdy, alert young men with serious expressions.

In fact the key defense in filtering was not the guards or the search, identikits, previous intelligence, curious travel and flight plan, or general behavior—but the use of *profiles.* These, a series of checks that could be used by any airline personnel, could narrow down the search for potential hijackers. In 1969 a group under Dr. Evan W. Pickeral identified thirty-five behavioral characteristics common to previous hijackers and ran a profile test with boarding passengers for Eastern Airlines. The FAA decided not to order the airlines to use the results. Eventually the airlines, rather than the FAA, accepted the validity of the profiles. Soon, however, one of the greatest deterrents to the psychopathic hijacker in America proved to be a relatively simple profile developed by Dr. David Hubbard, a Dallas psychiatrist. When used by airline personnel, it virtually eliminated the psychopath at the check-in gate. Unless a psychopath wanted to shoot his way onto the plane, a major threat was removed, not by guards, or by inspection and scanning, but by the knowledge of *exactly* what kind of person was a threat. For example in the United States, over ninety-five percent of the hijackers were men between the ages of eighteen and forty-five who had purchased one-way tickets in cash. Thus, it was possible for the most part to ignore little old ladies, or businessmen using American Express cards—nearly everyone else, in fact. In time nearly all flight personnel also went through training in order to recognize the potential psychotic hijackers and to understand how to take remedial steps. If on board a man announces that he has a gun and must be taken to see the flight captain, a stewardess can end the "hijack" fifty percent of the time by asking to see the gun, since half of the hijackers have no gun. More than any other technical means, the filtering process reduced the incidence of hijacking—almost totally eliminating psychopaths.

Elsewhere you could board a flight—in the Persian Gulf, for instance—with no inspection, change flights at Athens without additional inspection by using the transit lounge, and arrive over the Adriatic with a valise filled with weapons—and it was done. The filtering, however, when properly organized, regularly produced a collection of long knives, small revolvers, and drugs, and also weeded out many of the disturbed. But only occasionally did it turn up a real terrorist. A second line of defense was created in America and elsewhere with air marshals or security guards on the ground. In some cases, notably the American, their very presence, widely acclaimed after 1968, was assumed to be a deterrent. Very few American sky marshals became involved in midair incidents, but the Israelis put men aboard El Al planes with orders to shoot, not matter what the risk, which is how Leila Khaled's second hijacking attempt ended at Heathrow instead of Dawson Field. Those countries with lax airport control increasingly assured that someone's flight, maybe theirs, would end at Entebbe or Aden or Tripoli. While most nations were concerned with airport security, few wanted shoot-outs at 35,000 feet; and therefore, air guards did not become standard practice. On the ground, however, most international airports were guarded with varying degrees of efficiency. At Rome's Fumincino, after a fedayeen attack that killed thirty-one tourists on December 17, 1973, soldiers armed with submachine guns were stationed throughout the airport and on the tarmac. Given their general demeanor, they appeared likely to be a threat to the innocent in the event of subsequent attack. Other airport guards, especially at American airports, appeared as often as not to be superannuated rural police. Still it was thought that like the sky marshals, their very presence would be a deterrent.

Certainly, after each massacre the possibility of the next one concentrated minds on *contingency planning.* People began to ask, "What if?" In January 1974, the result of these considerations became apparent when a "mixture" of intelligence indicated that Arabs were going to use surface-to-air missiles to shoot down airliners. At Paris' Orly Airport seven hundred extra police were kept on twenty-four-hour alert. In Brussels armored cars were driven alongside all El Al aircraft, right out to the runways, and police with submachine guns were stationed in airport buildings. The major West German airports, already noted for some of the most stringent security, were assigned more men. Madrid and Athens stepped up security, and at Heathrow, British troops moved in with tanks and personnel carriers in a full alert. Even at Rome, everyone's choice as the most porous European international airport, the number of security men on each shift was increased from 250 to 400. Nothing happened. Perhaps there was nothing to happen. But the response had been heartening to those concerned—the long litany of massacre and disaster had not been futile. If this time intelligence had been in error, it might not be the next time. And there had been preintelligence before—rumors of an attack on an Arab embassy in an Arab country before Khartoum, or the arrival of fedayeen at the flight check-out desk, only to find security personnel waiting for them. And intelligence was the best deterrent of all.

One of the most sought after and most elusive of deterrents was an agreement on *no-sanctuary.* While hijacked airplanes were usually regarded as a poison parcel, still, ultimately, most hijackers could find sanctuary even if it meant brief imprisonment. The most effective agreement was the Cuban-United States antihijacking accord of February 1973, reached in negotiations conducted through the Swiss government. Havana had by 1973 recognized that most, if not all, of the American hijackers landing in Cuba were either criminals or psychopaths, not revolutionaries. Once the agreement went into effect, a criminal no longer had safe haven but rather assured arrest. Together with the psychological profile, the no-sanctuary agreement brought an end to the Cuban hijacking connection.

The great reduction in incidents by criminals and psychopaths left the revolutionary as a permanent and most dangerous threat. Some revolutionaries had friends in Libya or South Yemen or Uganda, or were too hot to prosecute without opening up a new cycle of hijackings, and were released. Neither the newly suggested international covenants and agreements nor specific hard-line national policies nor suggestions of punitive action by nations or pilots' associations solved the problem of sanctuary. The problems of effective legal punishment as a deterrent thus proved most difficult. Air piracy was obviously a crime, even if some nations did not have specific legislation outlawing it; and most hijackers could be prosecuted under appropriate statutes. Many hijackers were in fact arrested, prosecuted, convicted, and imprisoned.

For a variety of reasons, this chain simply became a filter to freedom. Some suspects were deported before the planned operations. Others, once arrested, were released on various technicalities. Some were not prosecuted and others not convicted. Even when convicted, few served their sentences. This by no means dissuaded those who advocated strict international laws with certain prosecution or extradition. Yet the weight of evidence indicated that a revolutionary hijacker, even if caught—except in a few cases, such as Israel, obviously—could anticipate only *pro forma* legal penalties and even then might hope for a violent rescue. It was this record of punishment that revealed as much as anything the limitations of deterrence. Any prospective hijacker knew that taking on El Al was a major and difficult operation filled with risk, but that elsewhere seizure of a flight was not difficult, the discovery of sanctuary possible, and punishment improbable. Whether certainty of punishment in an uncertain world is actually a deterrent remains a moot point. As the Croatians indicated, someone, someplace, always seems willing to take the risk, seize the plane, and begin a revolutionary hegira with hundreds of hostages.

Once a plane had been hijacked, the immediate need was for centralized control to accumulate intelligence, communicate with the involved, and make any ultimate tactical or strategic decisions. The problem was that in any international hijacking, such control was difficult to fashion, despite considerable technological assets. Even when a special headquarters was established and staffed with the proper experts, as had been the case with the FAA control during the Croatian hijacking, the vital decisions were often made elsewhere by other nationals with other priorities and prejudices—individuals who often had little touch with the unfolding discussions.

Even if a control center exists, the exploitation of available talent may prove impossible. Ultimately, of course, unless the hijacked airplane reaches a friendly sanctuary—and sometimes even then—someone on the ground is the negotiator of last resort, who directs the discussions, seeks compromise or concession, and has the final word and the authority to ensure the implementation of any accord. Some states have ready at hand both skills and strategies. But if the incident is spectacular, authority nearly always creeps upward, leaving behind the special skills and agreed upon strategies. Premiers and presidents end up managing the crisis, assuming that it can be managed at all. If matters work out as they did in Paris with the Croatians, there is generalized delight and mutual esteem; if not, as in Munich and elsewhere, acrimonious recriminations begin, either in public or in private. Still, the raw figures on hijacking have encouraged the public, usually unaware of the differing hurdles placed before the criminals, the psychopaths, and the political revolutionaries. In the United States the result was impressive: thirty-two attempts, sixteen successful in 1972; down to a couple in 1973; and subsequently almost dwindling off entirely—thus, until the Croatian incident, many assumed the threat no longer existed. In any event, filtering had been a notable success. Another, equally visible success in antiterrorist tactics was the growing capacity to save hostages.

The immediate response to hostage situations varied with national perceptions and professional training. It was, as with hijacking, not immediately apparent that there were different kinds of hostage-takers and different kinds of hostage situations. Deciding what to do when confronted with an unfamiliar phenomenon, a paucity of expert advice, and the pressure of time and politics produced an uneven record. Basically, as with hijacking, the actors can be divided into three groups, often overlapping, acting for different motives and with different priorities: criminals, psychopaths, and revolutionaries.

Criminals seize a hostage in order to bargain for gain, as in the traditional kidnapping, or in desperation during the course of a crime, in order to bargain an escape. The psychopath acts out of frustration in a fantasy world. Revolutionaries may seize and hold a symbolic victim to prove their capabilities, embarrass the government, and create a state of tension. They may trade their hostages, as did Brazilian MI-8 revolutionaries when they exchanged the American Ambassador Elbrick for fifteen Brazilian political prisoners, or they may not really anticipate receiving an exchange, so that the negotiations at a distance become the message, not a medium—any concession being a bonus. In some hostage situations, however, like Munich and Khartoum—situations which have been choreographed in advance to assure a confrontation, with a part written in for the security forces, the terrorists do not want to "escape" but to seize the stage for their drama. Munich was, thus, more successful as a drama because it failed than if everyone had flown off to Algeria or Libya. Occasionally, of course, revolutionaries find themselves concerned in an unplanned confrontation. In this they resemble criminals who find the bank they intended to rob surrounded with the police and use the customers as hostages.

As for the psychotic hostage-taker, he ordinarily is acting on the frustrations of the moment, threatening to kill his baby daughter or his wife or his banker unless "something" is done. On occasion he may mimic the revolutionary-staged confrontation, when the incident becomes an opportunity for catharsis—therapy by kidnapping. With criminals especially, and usually with psychopaths, the traditional response of law enforcement officers had been to go in after the culprits, risking, of course, the hostages and at times their own men. In the United States the traditional response was exemplified by the Los Angeles police shoot-out with the Symbionese Liberation Army. The new SWAT team was deployed, the appropriate warning given, and the battle undertaken—without knowing whether Patty Hearst was inside the house. And sometimes the Israelis have sent in troops, as in the case of Ma'alat, despite the risks to the children in the school. But after the Munich massacre responsible officials increasingly began to explore alternate means of response to terror.

In New York City this led to Schlossberg's course and to the Detective Bureau Hostage Negotiation Team under Lieutenant Frank Bolz. The first experience was the confrontation at John and Al's Sporting Goods store in Brooklyn, where the officer in charge turned down a consulting psychiatrist's advice to go in with gas—(given the risk of fire in a store full of ammunition, this was an argument for a holocaust)—and applied Schlossberg's methods. The tactic worked, and hostage negotiation was added to the police arsenal. A further elaboration of the negotiations method was suggested by Dr. Irving Goldaber, who set up a police training course that sought to strengthen the ties between terrorist and negotiator—in effect undertaking intense therapy during the course of negotiations so as to build on growing mutual confidence. This might in theory have seemed wishy-washy to pragmatic police officers. But in practice, hostage-bargaining worked, and waiting paid off in saved lives—lives of hostages and police, not to mention terrorists.

Gradually, a highly detailed picture of hostage and "terrorist" behavior began to emerge. The hostage, it was found, usually goes through three stages: (1) the trauma of seizure—the violence, the absolute loss of freedom, one minute a successful banker, the next unable to go to the toilet without permission; (2) a period of alienation from legitimate authorities, seen as negligent and ineffectual, and an identification with the terrorists as the only effective authority—a stage that usually lasts somewhat beyond release and accounts for the transformation of Patty Hearst into Tania and the kind words about the Croatians from the passengers after their release; and (3) a long after-shock, with various forms of anxiety and fear. In the case of the terrorists, each type must be approached by negotiators differently: (1) the criminal's first priority is his future, and bargaining can stress self-interest; (2) the psychopath, who may be largely rational in planning and execution of the incident, acts out of frustration based on reality or fantasy, and therefore the bargaining focus must be on easing that frustration through establishing a mutuality of concern; (3) the political terrorist can most easily be approached politically—self-interest exists but may be low, and psychic manipulation may be possible but not certain. In all cases negotiation, establishment of trust, reasonable discussion, the gradual passage of time, and the exhaustion caused by tension tend to lead to some sort of accommodation with the criminal, with the psychopath, and sometimes with the revolutionary. If the hostages survive the first few violent minutes, the chance of success is good, especially with psychotics, and even with revolutionaries.

In order to maintain the momentum of negotiations, a wide variety of tactics and techniques have evolved, not all universally accepted, and often grounded in research that is less than convincing. Psychiatrists and psychologists have constructed rationales for such manipulative bargaining, but many police officers seem to cope adequately on their own. Thus Dr. Frederick J. Hacker, a psychiatrist and psychoanalyst who is Director of the Institute for Conflict Research in Vienna, has given a detailed account of his part in the negotiations with two Arab fedayeen who in 1973 had seized Russian Jewish refugees as hostages. Hacker's judgments appear sound, sensible, pragmatic, and unrelated to his scholarly qualifications other than his long practice at talking with strange people. The result of the Austrian negotiations brought about the release of the refugees, safe passage for the terrorists, and the shutting down of the facilities used in the transit by the Russian Jews. It also brought indignant criticism from Israel and the United States—charges of concession to blackmail—and support from those who noted that no lives had been lost. The technique, however, was no more or less "scientific" than most negotiations with or without the presence of guns. Hostage-bargaining was simply a matter of distinguishing the type of "terrorist" one is confronting, establishing the ground rules and then trust, and hoping to the end that force will not be needed. While there is increasing recognition by security forces that hostage-bargaining works, especially with psychopaths, there is not yet agreement on how much bargaining should be done, how many—if any—real concessions should be made, and how long the process should take. Still, the Pavlovian shootout now has few takers, even within conservative police forces.

The best that can be done not only in hostage situations but also during all terrorist incidents is to deploy the appropriate forces and techniques and try with contingency planning and international cooperation to cope as best as possible. For an outraged and indignant public, too often this has not been enough. The improvement in tactics and techniques and technologies has not always been sufficiently visible—even when clearly effective, as in the case of the Cuban no-sanctuary agreement or the hijacker profile. What the public wants is *something* visible to be done. The demand for actions has often focused on legislation in states where there has been an assumed identity between law, order, and justice. Punitive measures, including the death penalty, have been urged. Greater powers of investigation and intelligence for law enforcement agencies have been demanded, and punishment by international action of those who provided sanctuary has been called for. The democratic public demands laws, albeit mainly punitive, in the belief that without order there can be no justice.

CHAPTER

SOCIETY AT PERIL: LAW, ORDER, AND JUSTICE

We must not make a scarecrow of the law
Setting it up to fear the birds of prey,
And let it keep one shape, till custom make it
Their perch and not their terror.

Richard III,
Shakespeare

After each terrorist spectacular there were throngs who sought to transform the scarecrow shape of the law to frighten off the violent birds of prey. In December 1975, a no-warning bomb detonated in New York City's LaGuardia Airport and killed eleven people—no one ever took responsibility for the blast, and the police have not yet discovered the culprits. Public opinion was outraged. Immediately there was a demand for action. Representative Paul Rogers, a Democrat from Florida, announced that he would file legislation requiring the death penalty for anyone convicted of bombing public facilities:

> If we provide severe penalties and if persons convicted can be sentenced quickly, this will be brought to a halt. . . . People are angry about this—they don't want it started here. . . . To let this get started really could be . . . disastrous. . . . a very positive step . . . can be taken to show the outrage of the nation. . . . There'll be some

literally insane [people] that it won't stop but many I think it will. I think it will stop fringe groups. . . . This nation is not going to tolerate actions of that type.[1]

In a very real sense Representative Rogers' advocacy of the death penalty represented not a prospective solution but an aspect of the problem of response to terrorism. Rogers, like his constituents and most Americans, was outraged and indignant and wanted *something* done. The death penalty sprang to mind: at least *something* will have been done, and it might even be a deterrent. All the arguments have been worn thin, all the endless studies and statistics and correlations swept into the present fray and still no one really knows if the death penalty "works." But twice after the advent of the IRA no-warning bombs in England, Commons was divided on the reinstatement of the death penalty.

What has compounded public outrage is not only that terrorists have slaughtered the innocent, unsuccessfully covering the massacre with a fig leaf of revolutionary rhetoric, but that time after time they have escaped punishment and have gone free to plot and to plan again and, in Leila Khaled's case, to hijack a second plane. Hardly anyone outside the five Croatians and those the Israelis captured had been punished. For most terrorists it was the same story—released, freed by friends, no prosecution, or sentenced but released. Okamoto ended up in an Israeli jail and his colleagues dead, but the survivors of Munich and the Khartoum group went free. The Croatian assassins in Stockholm ended up at the end of the hijack-rescue in Madrid with *pro forma* sentences and an early release.

Thus each new outrage engendered a double load of wrath. One of the reasons that the Israeli raid at Entebbe was greeted with such enthusiasm was that at last the "evildoers" had been punished; even in distant Uganda they had not been beyond swift vengeance. Whether such vengeance was

[1] *Palm Beach Post*, 1 January 1976.

legal or proper was uncertain, but its effectiveness was never questioned. In any case, in the West, whether on the editorial pages, in the language of television commentary, or within the various journals of opinion, the immediate response to terrorism from the first was outraged indignation and a call for action.

The law, then, like the police or the government or even the academic community, is *supposed* to act. And if the death penalty would not be effective then the lawyers must devise a legal avenue that will give the police greater power of search and seizure, or introduce internment, or end the right of habeas corpus, or allow electronic intelligence, or punish countries that offer hijackers sanctuary, or force an appropriate resolution through the United Nations. But *something* must be done. And hovering there, just behind the indignation and the demand for action, is the unstated implication, "If you don't do something we will."

Within the academic legal community there are the same ideological divisions to be found elsewhere. On one hand, there are those who focus on state terror, on the legitimate aspirations of guerrillas or freedom fighters, and on the dangers of official repression; on the other hand are those who contend that there is no excuse for the massacre of innocents or the hijacking of airplanes or the kidnapping of ambassadors. Both sides, however, retain a faith in the law—in the need to protect human rights with legislation or to prevent hijacking through international covenants or to punish those guilty of massacre through the agency of a new world court. Both see terror as a legal problem.

In fact the number of pages churned out by the profession which analyze the terrorist phenomena, suggest legal remedies, and comment on the views of professional colleagues probably equals the total of all the other published academic terrorist material. In part this reflects the real possibility of shaping legislation or legal institutions and in part the real capacity to review previous efforts to confront terror with the law. Within the legal profession there persists a belief, so far demonstrably false, that the structure of laws will either deter the determined or assure their ultimate punishment and that such laws will guarantee not only order but also justice. Louis M. Bloomfield and Gerald F. Fitzgerald, in their *Analysis of the United Nations Convention for Protected Persons,* described the convention as "one of the most remarkable achievements of the United Nations" which "will be assured of an important and effective role in legal history even though it does not result in full elimination of attacks against internationally protected persons."[2] The assumption is based on no visible evidence, but rather on faith in the profession and the power of the law. But George Habash or Carlos or Okamoto answer to a higher law.

In fact those in positions of power who have authorized votes in favor of such conventions, called on legislative bodies to pass more effective and vindictive legislation, and expressed their horror at the futile and mindless massacre of innocents have been known to evade or translate the law to political advantage when necessity demands. Thus the suspected mastermind of the Munich massacre, Abu Daoud, arrested in Paris by security forces, was subsequently shipped off to Algeria, all quite legally. The British found that it was quite legal for them to send Leila Khaled back to her colleagues of the PFLP, since after the failed hijacking of the El Al jet that ended on the ground at Heathrow she was on British ground "irregularly"—had not really been admitted into the country.

Still it is contended that while man has not managed to outlaw war or torture or brutality, despite legislation, at least the law can do no harm. This is, of course, not true either; for while the lawyers, with rare exceptions, advocate international remedies—covenants, courts, treaties—the govern-

[2] Louis M. Bloomfield and Gerald F. Fitzgerald, *Crimes against Internationally Protected Persons: Prevention and Punishment, An Analysis of the UN Convention* (New York: Praeger, 1975), p. 147.

ments under threat insist on greater power and wider scope for the security forces, seeking to protect an open society by closing it down. In general the legislation fashioned in response to terrorism to achieve order through law can be examined on three levels: international, regional, and national.

Order Through Law

International Legislation

Attempts both to define terrorism for legal purpose and then to construct effective international legislation stretch back for over a century, although for practical purposes the first real step came at the First International Congress of Penal Law, held at Brussels from July 26 to July 29, 1926. At later Congresses the question of political crime was considered; and in 1935 the term "terrorism" was expressly used for the first time. By then there was additional movement following a flurry of spectacular assassinations—Romanian Premier Ion G. Duca, Austrian Chancellor Engelbert Dolfuss, and then, simultaneously, King Alexander of Yugoslavia and French Foreign Minister Louis Barthou, on October 9, 1934. Immediately after the later incident, the French government dispatched a memorandum to the League of Nations calling for an international convention for the suppression of political terrorism. At the League an eleven-member committee by 1937 worked out two conventions, one for the repression of terrorism and the other for the establishemnt of an International Criminal Court. And for the most part, the matter rested there. Only three nations ratified the first convention; none ratified the second. The League's prestige had soon been dissipated, and the furor of the world war ended any hope of an international agreement.

A variety of international responses to war and conflict,

beginning with the Hague Regulations of 1907 and continuing through the Geneva Conventions of 1949, had little direct relevance to terrorism. And even the status of "the guerrilla" is still a matter for dispute and interpretation, although in May 1977 the 1949 Convention was adjusted to legalize guerrillas and partisans. It was not until after the Munich massacre that a specific attempt was launched at the United Nations to deal directly with terrorism. On September 25, 1972, the United States submitted to the General Assembly a draft of a convention for the prevention and punishment of certain acts of international terrorism. Previously, a similar convention had been approved and signed—on February 2, 1971—by thirteen members of the Organization of American States. What seemed to the United States and most Western countries so pressing an issue so obviously in need of an international response did not so seem to a great many states of the Third and Fourth World. While deploring murder and massacre, these states had other priorities, noted state terrorism as well, and privately feared that such an international convention was directed against the legitimate struggles of national liberation. The Libyan representative spoke for many when he suggested that the convention was a maneuver aimed at thwarting the struggle of the people against colonialism. A compromise failed, and an ad hoc committee was later formed, but without effect. Subsequently, the General Assembly could not find time to discuss terrorism. In September 1976, the West German government submitted a treaty to the General Assembly in which Foreign Minister Hans-Dietrich Genscher scrupulously avoided using the word terrorism:

> New forms of illegal force, such as the seizing of hostages, are developing into a world-wide plague. None of the 500 million passengers travelling on airlines every year can be sure not to be among the next victims.[3]

[3] *New York Times*, 29 September 1976.

Whether the new maneuver would ease the other delegates remained uncertain, even with Secretary General Kurt Waldheim's support for such action. Elsewhere, however, progress of a sort had been made on approaching the spectrum of terrorism bit by bit. In fact, on December 14, 1973, the General Assembly adopted without objection the Convention on the Prevention and Punishment of Crimes against Internationally Protected Persons, including Diplomatic Agents, but few governments rushed to ratify the convention.

Outside the United Nations the world-wide spread of airliner hijacking has led to international responses to deal with this threat to an international industry. On September 14, 1963, in Tokyo the International Civil Aviation Organization (ICAO) adopted the Convention on Offenses and Certain Other Acts committed on Board Aircraft, which contained no real threat of punishment to deter hijackers. At the Hague on December 16, 1970, a second Convention for the Suppression of Unlawful Seizure of Aircraft was signed by delegates from seventy-four states. The signatory states now, if they did not extradite the hijacker, were obliged without exception to prosecute the offender. A further ICAO Convention for the Suppression of Unlawful Acts Against the Safety of Civil Aviation was signed at Montreal on September 23, 1973. At this point the tide shifted. At the ICAO conference in Rome between August 28 and September 21, 1973, the American attempt to get stronger penalties approved failed. First, to search for fedayeen, the Israelis had forced down a Lebanese commercial flight on August 10, 1973—two weeks before the conference opened. The Arab delegations arrived at the conference with the demand that Israel be expelled from the ICAO. Neither they nor many of the other delegates favored stronger sanctions. The Americans, realizing their isolation and the prospect that Israel would be expelled, encouraged the conference's collapse in order to salvage Israeli membership by a procedural maneuver. The international efforts to protect civil aviation had gone further and in greater detail than most similar efforts, but basically, the more stringent the regulations and penalties the more reservations expressed by the membership—a membership too diverse to permit consensus.

Regional Legislation

In theory at least, states with similar institutions or at least geographic ties might more easily find common ground. Thus the Organization of American States that approved the terrorism convention in February 1971 would appear to be a shining example of regional cooperation—except that Cuba was not a member and Argentina, Brazil, Ecuador, and Paraguay for varying reasons were not present, and Bolivia and Peru had abstained, and Chile voted against the convention. The result was really not an effective convention. Given the existence of various bilateral Latin American rivalries, probably no convention was possible, or else it was likely that any accepted convention would be ineffectual. At least it was an exercise that might in time lead to cooperation among some of the member states. What *did* make regional agreements work was a mutuality of interest—thus the Feburary 1973 bilateral Cuban-American hijacking accord was the single most successful regional effort. As soon as the mutuality of interest began to spread thin with the addition of further signatories, conventions had a tendency to become merely cosmetic.

In July 1976, the nine Common Market countries, under the urging of German Chancellor Helmut Schmidt, began to draft an antiterrorist treaty. A broader convention was subsequently undertaken by the Council of Europe. In 1976 a draft convention was proposed. There was a great deal of adjustment and rewriting—particularly because of Irish concern about the provisions relating to extradition that might force them to hand over Northern Irish Republicans on the run in the south to the British Army—hardly a pleasing political prospect. The convention was at last accepted by the

Council of Europe in November 1976 at Strassbourg, although Ireland and Malta failed to sign it. It was to go into force when it had been ratified by at least three countries in the nineteen-member Council. The convention lists as acts of terrorism hijacking; taking hostages; abduction; sequestration; the use of bombs, grenades, and firearms; any infringement on the rights of diplomats; and damage to public buildings. None of these acts would any longer be considered a breach of law inspired by political motives, and the government involved would either have to extradite or try the terrorists. Seemingly there would no longer be sanctuary for a terrorist anyplace in Europe.[4]

On January 3, 1977, the director of the Paris office of the Palestine Liberation Organization, Mahmoud Saleh, was assassinated. Two days later the PLO asked the French consulate in Beirut to issue a visa for Youssif Hanna Raji to attend the funeral. On January 13, French Premier Raymond Barre indicated that the French police on Friday, January 7, while investigating the assassination of Mahmoud Saleh, began to have doubts about the real identity of Youssif Hanna Raji. They discovered him to be Abu Daoud, of the Munich massacre. At 6:30 P.M. on January 7, the German police informed the French police that they were requesting the German judicial authorities to issue an arrest warrant for Abu Daoud. An hour later the French brought in Abu Daoud. A German telegram arrived for the French Minister of the Interior announcing a forthcoming request for extradition. Abu Daoud was detained. However, when he was brought before the *chambre d'accusation* it was discovered that confirmation of the German request had not been forthcoming. The German Embassy provided no further information. There were thus no grounds to justify Daoud's continued detention nor grounds to accept an Israeli request for extradition. The Minister of the Interior issued a deportation order immediately.

[4] Excluding Malta and Ireland, the West German constitution prohibited extradition of German citizens.

On January 11, Abu Daoud was escorted to Orly Airport, where he boarded a flight to Algiers.

When the news broke that Abu Daoud had been released, there was a week-long, worldwide storm of protest. The Israelis recalled their ambassador to France. Newspapers and television commentators castigated the French for freeing the butcher of Munich. President Giscard D'Estaing, and other French spokesmen, were in turn outraged that they were attacked simply for following legal procedure. The detailed explanations of the function of the French legal system did not convince critics—who assumed that *some* way to hold Abu Daoud might have been found if it had been to French advantage. Clearly, the signature on the European convention had not been relevant to the Daoud case; and under different conditions in any of the signatory countries the convention would perhaps not be relevant when weighed in the scale of justice against political considerations.

National Legislation

Although nearly every state has evolved a policy toward international efforts to legislate against terrorism, often the most effective or the only effective legislation has dealt with internal matters, by increasing initiating or increasing penalties for specific acts or restricting the civil liberties of some in the name of order for everyone or extending the role of various security organizations. Generally, the political demand for such legislation has arisen from a perceived threat, and often the enacted statutes add little to the law's arsenal but much to the satisfaction of an outraged public. In Germany, for example, as the student unrest peaked in 1972 with the murder of American soldiers by the Baader-Meinhof group, a law known as the Radicals' Ordinance stipulated that those seeking civil service jobs undergo political screening and be rejected if they have been affiliated with "extremist" groups. Over half a million applicants have been interrogated to date, and four hundred have been rejected. Willy

Brandt called the interrogations a grotesque failure (although he was Social Democratic Chancellor at the time the law passed), and there were complaints in Germany and Europe that the Teutonic authoritarian heritage was reemerging.

Yet much the same process occurred elsewhere. In Italy under the sponsorship of a Socialist Minister, an emergency measure—one that even he admitted might have gone too far—was enacted to counter the rise of neo-Fascist Black Order and ultra-left *Brigate Rosse* violence. In Britain after the Provisional IRA no-warning bombs went off in Birmingham, Westminster passed a Prevention of Terrorism Act that permitted detention of suspected terrorists for up to seven days with no guaranteed access by lawyers.

Even a small and apparently immune nation like Sweden has felt it necessary to enact a law as a response to terrorist provocation. The first serious terrorist incident in Sweden occurred on April 7, 1970, when the two Croatians killed the Yugoslavian Ambassador Vladimir Rolovic. The two were arrested, tried under the existing criminal code, convicted, and sentenced to long prison terms. The following year they were freed by the three other Croatians who hijacked an SAS DC-9 and negotiated their release plus the release of four other Croatian prisoners and the payment of the 500,000 kroner ransom. The Croatians were flown to Madrid, arrested, tried, and sentenced by Spanish authorities, but then quietly released in 1975. In response to this hijacking, a Swedish Commission for the Prevention of Political Acts of Violence prepared antiterrorist legislation. The major aim of the law, which went into force in May 1973, was the control of foreigners in Sweden: requirements for entry visas were tightened and new grounds for expulsion from the country and/or prohibited entry were established; court-authorized electronic surveillance was permitted, along with postal and telegraph control and extended searches. In October 1975, new airport control, including search procedures, was introduced. Yet this period saw continued terror spectaculars in Sweden, including the seizure of the West German Embassy in Stockholm in April 1975 by six members of the Baader-Meinhof Group that resulted in the death of two embassy officials and one terrorist and the bombing of the building.

As for the United States, any summary of its legal response to terrorism is hampered by the vast number of governmental and law enforcement agencies involved. In the whole of Great Britain, for example, there are only fifty-one police forces; while in the United States, at least three thousand different forces have SWAT teams, and of course, other thousands do not. On the most basic level, the first attempt to enact a uniform national penal code is still pending before Congress. In the meantime, the federal government has relatively rarely been involved in enacting legislation dealing with crimes arising from terrorist actions. In September 1961, a law was passed instituting penalties for air piracy that ranged from death to not less than twenty years imprisonment. After the Supreme Court decision of 1972 holding that the death penalty was unconstitutional because it was capriciously imposed, a new law was enacted that sought to evade the problem of "capriciousness" by reimposing the death penalty, subject to special hearings and assurance by a jury that there were no mitigating circumstances in hijacking cases involving death. The Croatians evaded the death penalty by arguing a variety of mitigating circumstances. For the most part United States federal authorities have made do with existing laws; although on April 30, 1970, a statute was passed making the murder of foreign officials and diplomats a federal crime.

A far more troublesome American problem has been how to define the scope permitted to law enforcement and intelligence authorities, especially in the long and bitter fallout after the Watergate scandals that revealed that the FBI had not only been involved in illegal acts but also had chosen targets that were hardly serious threats to American society—student protestors, the women's liberation movement, the Trotskyite party, black civil rights figures, and Arab information officers. The CIA, too, appeared to have been involved in domestic activity, seemingly in violation of the original charter. That is why, whatever their past misdeeds and ill-considered enthusiasms, officials of federal agencies and of local police forces today fear that their powers of investigation will be too severely limited in the continuing backlash. On the other hand, except for the Puerto Ricans, the United States has spawned no native terrorists who seek to kill. Most radicals have been content with symbolic bombs that have not fueled political pressure for emergency legislation.

Order without Law: The Court of Last Resort

Many democratic states that have not had the advantage of America's immunity from native revolutionaries have not only been stages for others' spectaculars but have also suffered domestic outrages. Then the demand for emergency legislation has resulted in legislation but not always order; the *Brigate Rosse* and Black Order in Italy, the Second of June Movement and the Baader-Meinhof Group in Germany, the South Moluccans in the Netherlands, the Provisional IRA in the United Kingdom and always the Croatians continue to disrupt Western Europe.

When the limits of the law have been exhausted and there is still neither order nor justice, the temptation for the state is to mimic the practices of the terrorist groups—especially if the conflict takes place at considerable distance from the capital. But most colonial wars are over and today the conflict is most likely to be waged in the streets of the capital, where gunmen shoot down the servants of the state. In Uruguay the army finally closed down Latin America's premier open society, institutionalized torture and brutal repression, and thus crushed the Tupamaros. In Israel the Wrath of God, a state institution, borrows the tactics of terror wielded by the fedayeen. And at home Jerusalem still deploys the tactics and techniques of repression inherited from the emergency legislation of the British Mandate.

The spectrum of state torture, even in democratic states, runs from brutal interrogation through those who murder on demand. During the Algerian war the French Red Hand sought out and killed in Germany and France those agents who were supplying arms to the FLN rebels, just as the Israelis seek out and kill the fedayeen. In Northern Ireland Britain has been accused of introducing a variety of practices most readily defined as torture—except by the investigating judge, who felt that if the soldiers did not enjoy what they were doing it could not be called torture. The British underwrote the bank robberies in Dublin, and the Americans supported insurrections by the Kurds and invasions by the Cubans. According to former President Nixon, a president has the right to do wrong; and not just the right, but also the responsibility. Those who insist on order have lost patience with the forms and substance of the law. If the law cannot achieve order, then additional and more stringent steps must be taken outside the ineffectual law. After 1971 Northern Ireland, for example, became the arena of a classic guerrilla campaign fought largely within legal rules. It was these "rules" that the ultraloyalists felt were permitting the IRA to operate. Beginning without notice in 1972, various ill-organized loyalist groups—the Ulster Volunteer Force or the Red Hand Commandos or the Ulster Defense Association—began a random campaign against Catholics, which in turn led to tit-for-tat murders, no-warning bombs, and rural massacres.

In the case of Ulster, while hardly maintaining an evenhanded approach to both the IRA and the loyalists (except for a few individual soldiers), the British did not tolerate, much less support, the vigilantes' murder campaign. Like the Ku Klux Klan in America after the Civil War, the vigilantes were killing in support of a society that the state either opposed or would not protect. But there have been times when some states have been delighted to have some of the more unsavory defense chores performed by unrecognized allies. There were those in the government of Weimar Germany who saw virtues in the Nazi attacks on Communist cadres. In authoritarian states official and unofficial vigilantes are even more common. In contemporary Argentina it is difficult to discern the line between the military and police and those who torture and murder "suspects." Democracies, however, usually discover the private gunman only at moments of most desperate crisis. Indeed, his emergence from the shadows is in itself a warning that order has truly eroded, as is the case in Northern Ireland. There with justice nowhere to be found, with the law mocked, with order eroded, the fields and back lanes have bred vigilantes who seek to impose their violent dream on the province.

The Limits of Law and the Bounds of Justice

On a national level the law can make the policeman's life a happy one by extending his powers and prerogatives, but at the expense of an open society. Even in brutal, authoritarian societies there is usually *pro forma* recourse to the law. In democratic societies war measures, coercion laws, special powers and emergency legislation have had a long and not always happy history. Even when the necessary legislation does not exist, democratic leaders have somehow found the authority to intern aliens or authorize burglaries and murder. When legislation is passed in response to a spectacular provocation—a no-warning bomb, a dramatic assassination, urban riots—there is a tendency to maintain it on the books after the emergency has disappeared—just in case.

Yet such laws do not produce order. In the nineteenth century the number of British Coercion Acts passed to deal with the Irish was staggering, and yet the Irish kept on revolting. If, in the matter of repressive legislation, less is more, it is almost too late even in most liberal democracies to whittle away very much of the power of the state. In this respect the American experience is most unusual in that after Watergate, it is politically advantageous to limit the powers granted to the guardians of order. The present process has produced severe anguish in those who see the country in a daze of "reform," discarding the tools necessary to protect democracy.

In the case of the terrorist provocation, the whole variety of national-based legislation has been largely directed at raising the cost for the potential offender at a price in general liberty. In the equation ideally the first should be maximized and the second minimized. In legal matters, as long as existing criminal codes can be employed, there is no real need for further laws written to ensure order. Yet when juries will not bring verdicts, when justice is warped by extortion and fear, when conventional codes and procedures falter, most in authority insist on special powers. At times such legislation has been accompanied by acts that attempt to balance coercion with concession. In sum there is no ready answer and those answers readily supplied, like much else in the terrorist muddle, are more likely to be based on long-lived ideological positions and personal predilections than on pragmatism.

On an international and regional level the record of the law also leaves much to be desired in pragmatism. No one really believes that the Kellogg-Briand Pact ended wars. Yet the exercise at the United Nations and in various regional organizations has a considerable part to play in maintaining world order: first and foremost because such covenants can be seen by the outraged public as *something*—an act, an action, a response. More important, unlike national emergency legislation that may restrict liberties unnecessarily, the very ineffectuality of such international agreements means that the law is benign. Thus a covenant may not work at all as a "law" but may do wonders as a palliative. And the debate and concern arising from such exercises readily underlines for most governments their largely hypocritical posture on the evils of terrorism—the Abu Daoud affair, the endless free fedayeen affairs. It is more difficult, although hardly impossible, under the glare of television, with the ink not dry on an antiterrorist covenant, to act solely in the narrow national interest or for immediate political gain. Consequently, the activities of the international jurists and scholars have much to be said for them, even if the end-product will not work as intended. When laws do work, as did the Cuban-American agreement, it is because their enforcement is to mutual advantage. With over 150 states, it is not very reasonable to hope that any law would be to everyone's advantage. But it is not unreasonable to hope that the Council of Europe's covenant will work.

Finally, in terrorist matters the overwhelming burden of concern has been on the nature of maintaining order by deterrence, by recourse to extradition and certain punishment of recognized crimes, and by international legal cooperation in such matters. Only from time to time have there been reasoned efforts to suggest a course of response that would attempt to assure a justice denied, a denial that encourages the faithful to resort to violence. When arguments have appeared they have focused on various national liberation movements of the moment, whose defenders explain the necessity of responding to imperialism, racism, and state terror by the only tactics available to the weak. Here the problem is that the law's role is relatively minor, defining previously forbidden acts—guerrilla war—as now legitimate. Legal institutions are hardly in a position to impose justice; the Turkish army is still in Cyprus, despite overwhelming censure by the members of the United Nations. A further problem arises in that the aspirations of some —the Japanese Red Army or the Baader-Meinhof Group —seem beyond reasonable concession. Perhaps some day there might be a Palestinian state or, some day sooner, a Zimbabwe; but what justice does Carlos want, or Okamoto? Justice for all, then, is the ultimate and impossible dream of those who seek an orderly world under law. There will always be injustice, perceived or real.

If the new terrorism is not really a legal problem, this has not meant that the law is irrelevant to the maintenance of order in the West. The broad international convenants serve not only as a palliative but also in some cases as a deterrent. Regional and bilateral agreements do work. In 1977 Finland returned hijackers to Russia under a treaty obligation, and Sweden tried two other Soviet hijackers as a result of local law and international obligations. National law, however, presents the cruelest dilemma, for a balance must be reached between the real need of the security forces and the liberties of all. And a balance is most difficult for politicians to achieve when all too often the many demand both vengeance and the erosion of their own freedoms in the name of necessity.

CHAPTER

NATIONAL POLICIES: THE SECOND ROUND, 1973–1977

". . . violence is the one language the Western democracies can understand."
Ilich Ramirez Sanchez (Carlos—the Jackal), spoken in the air after the OPEC operation

Curiously or perhaps not so curiously, recourse to law to achieve an orderly world seemed about the only constant in all national policies—Libya was against hijacking, and even the grotesque Idi Amin in Uganda urged respect for the law. The Council of Europe, composed of a substantial portion of the world's democratic states, has managed a regional agreement. Elsewhere most democratic nations had backed the West German initiative in the United Nations. Thus there was a consensus that "terrorism" was everyone's problem and an agreed approach dealing with that problem was through legal covenants. This was true in the case of every democratic government—except when such covenants or agreements might prove cumbersome and thus had to be defined away. States still continued to offer sanctuary to approved terrorists, to release unwelcome prisoners, to negotiate if need be, and to concede to demands if driven. Yet, the covenants and proclamations were not solely a patina over pragmatism. The responsible often complied with the regulations, not always to their advantage. The twisting and turning in public to find appropriate loopholes became more difficult and often embarrassing. There was by no means an agreed policy among nations when threatened by terrorists, but the bounds of disagreement were recognized and, certainly among mature democratic states, the desirability, of a universal policy encoded in law was recognized.

Given the scope of the agreed analytical wisdom and the variegated nature of national priorities and traditions, there was no wonder that as the years passed after Munich 1972 no single satisfactory, effective response emerged from the repeated incidents. There was still disagreement on the nature and rationality of the terrorists and consequently on the most appropriate response.

On one side of the spectrum within the government or the law or the university were those who felt that concession was fruitless. For them the terrorist was largely irrational, seeking unnegotiable items chosen from a bloody shopping list. The terrorists might be in the service of other more reasonable people, but they—the gunmen—were beyond accommodation. The Baader-Meinhof Group, the Symbionese Liberation Army, and the Japanese Red Army were fanatics more easily explained by psychologists than political scientists. Even those who felt that many groups were composed of reasonable, if fanatical, members agreed that the espoused aspirations could never be achieved, that political compromise and substantive concessions were quite impossible.

In the middle were many who denied that all terrorists were either irrational or beyond redemption. Perhaps nothing could be done to or for the tiny Japanese Red Army, but a mix of concession and repression, recourse to adequate safeguards and patience, might work with other organizations that represented a political reality and could not be countered simply by defining the members as mad and refusing even to contemplate negotiations.

Further along the spectrum were those who insisted that

the most useful tools in countering the worldwide phenomena were accommodation, concession, understanding, and sympathy. Certainly the Palestinians performed violent and unsavory deeds; but they had a just or at least understandable cause, were desperate, driven men without other weapons. Even if concessions were impossible, at least the wealthy and secure, the great powers, could listen to the anguish of the driven, thus making the wretched of the earth less violent, if not content. Finally, there were some of course who did not really care what horrors their friends employed as long as the cause was pure; this was especially true of the fedayeen. Why was the Western world so concerned over hijacking and not over the regular Israeli strikes on the refugee camps? Why did a country like America assume the right to criticize others after Vietnam, after Hiroshima, after Dresden?

And so it went from repulsion at the dementia of psychopaths, a phenomenon of the times unrelated to politics, to enthusiasm for the sacrifice of patriotic freedom fighters against global evils. And clearly, in most part national policies reflected this analysis—shifting rapidly, however from enthusiasm through toleration to repression when terrorists showed up on the native doorstep. Sweden was transformed from advocate of Greek liberation—that is, armed Hellenic revolution—to an arena for German and Croatian gunmen and thereafter revealed less sympathy for national liberation and more concern about emergency legislation.

Even those countries, including Western democracies, that tended toward accommodation realized that certain safeguards beyond a pure heart were needed. Thus although national postures changed there was everywhere a movement toward increased security. The most obvious change was that while the terrorists could still shock and outrage the millions, they could no longer surprise anyone; for the terrorist drama was all but endemic in much of the world. Consequently, potential targets had prepared not only special forces or control centers but also often avowed strategies of response.

These Western strategies generally related not to a world view or a general theory of accommodation or retaliation but to guide rules for incidents. A few countries like Israel did, indeed, have a strategy based on a global view, a reading of their opponents, and a set list of political priorities. Others had a scale of values: protect the innocent no matter what. Several governments shifted posture and position because of personnel changes or political fortunes. The most significant change in the years after Munich, however, was the increase in technical and tactical options—hostage-bargaining skills, SWAT forces, background intelligence—that permitted a flexibility and adjusted response. At the same time that Western leaders could rely on the new techniques, they also sought at a more general level to achieve a legal consensus through laws, covenants, and understandings on as broad a level as possible. The most visible aspect of the Western response, however, was not the level of deterrence or the number of national and international agreements, but rather the reaction to terrorist provocation.

Retaliation

Israel, with its Wrath of God strike force, remained alone among democratic nations in seeking to respond to the threat by extending the scope of the conflict and seeking out and destroying avowed enemies wherever they might be. The policy soothed domestic anguish, fit neatly into Israeli practice, and reflected long-lived military and political assumptions concerning the nature of the real world and the Arab mind. And the Palestinians were particularly elusive opponents, without a solid base except for the refugee camps that over the years had regularly been the target of Israeli airstrikes.

The vulnerability of Israel to the strikes of nomadic revolutionaries and their friends was revealed once more on June 27, 1976, when a group of transnational terrorists seized an Air France jet airbus with 257 passengers on board on a Tel Aviv to Paris flight and directed the pilot to fly to Entebbe in Uganda. Three thousand miles from Israeli vengeance or French influence, the hijackers announced that five countries, including Israel, must release fifty-three revolutionary prisoners by July 1. On June 30, the hijackers released forty-seven hostages. The following day, Israel publicly agreed to negotiate, apparently breaking years of a no-negotiation policy. The hijackers of the PFLP, including two Germans, then released 101 passengers and extended the deadline. There was general jubilation in the Arab camp. The Israeli concessions, however, were only a cover for the airborne commando raid that hit Entebbe Airport on July 4. The commandos freed 103 hostages, sabotaged the Ugandan air force, and flew off to a tumultuous welcome at Ben Gurion Airport. Four Israelis, seven terrorists, and twenty Ugandans had been killed. One hostage had been left in a hospital and was subsequently murdered. There was in much of the Western world great acclaim—at last the men with the guns had been smashed down when they felt most secure. Any legal quibbles over the Israeli raid were brushed aside, while United Nations Secretary General Kurt Waldheim called on the world community to act urgently against hijacking.

The Entebbe raid was perhaps the single greatest success of the Israeli campaign of retaliation, not because the terrorists were punished or because it would be a deterrent to future gunmen, but rather because of its impact on Israeli morale. Depressed by scandals, by inflation, by the revelation of vulnerability and lack of leverage after the October War 1973, the public grasped Entebbe as a victory. And to a very large extent Israeli retaliation policy—one facet of a continuing war—was as concerned with public morale as with the Arab threat. There could be little doubt that the drama of Entebbe had transformed the Israeli public and had been received with delight in many parts of the world where terrorism was abhorred. The Americans or British or French might not choose retaliation, but they appreciated the Israeli response.

No Compromise

The base of the policy of no compromise is that concessions simply lead to further demands; to save a few lives now would put more lives in jeopardy in the future. Such a policy also gives the government involved an aura of firm, adamant control, strength in the face of adversity, uncompromising opposition to evil—unless, of course, the victims about to be sacrificed are too important and have too many friends or the terrorist threat if enacted would be too dreadful. Even the Israelis at times have contemplated concession, and Rabin's government at the time of Entebbe was under severe pressure to make some concessions. In any case the no-compromise position obviously is congenial, for neither politicians nor their public want to give concessions extorted by violent men.

The Croatian hijack from New York to Paris had given the French government the opportunity to display such an absolutist attitude. For Poniatowski and Giscard d'Estaing any negotiations were out of the question, a concession to blackmail. At De Gaulle Airport the French did not want to wait even to allow the Croatians to confirm that leaflets had been dropped. The answer to the TWA Captain's plea, "Tell me, please, what are we being killed for?" apparently was a mix of pique and impatience.

In contrast elsewhere the overwhelming national response to terrorist sieges was great patience, even when there was no intention of compromise. Thus in Ireland in November 1975 the Irish police patiently waged the thirty-six day siege before securing the release of the Dutch industrialist Dr. Tiede Herrema. Later, at the siege of Balcombe Street in London, the four IRA men came out after six days. At the start of the siege, Sir Robert Marks had announced that the IRA members were going nowhere but to prison, and the police simply waited them out for 138 hours. The fact of the matter is, however, that international confrontations complicate no-compromise strategies, for much depends on the control of negotiation—and not to negotiate at all may lead to disaster, as it did at Khartoum.

The Netherlands, like so many other European countries, became an arena and a target for the transnational terrorists. KLM planes were hijacked, Palestinian fedayeen and Japanese Red Army cadres launched operations in the Netherlands, and hijackers thousands of miles away presented non-negotiable demands. At first the Dutch generally opted for conciliation and accommodation. For example, on September 12, 1974, three members of the Japanese Red Army seized part of the French embassy in The Hague, capturing the French ambassador and other hostages. Remembering Lod, the Dutch government persuaded the French to release the Japanese prisoners as the Red Army people demanded. A French jet with a Dutch crew flew the four Japanese to Damascus.

On November 22, 1974, a British jet was hijacked by the fedayeen, who then demanded that the Dutch release two fedayeen who had been arrested in March and sentenced in June to five-year terms for hijacking. The Dutch agreed. On November 25, three days later, a KLM 747 jumbo jet was hijacked, giving the fedayeen 247 hostages. The KLM jet was forced on a long odyssey—Damascus, Nicosia, Tripoli, Malta, and finally Dubai. The passengers were released in dribs and drabs until the last eleven captives left the plane at Dubai on November 28, after the Dutch government had made a solemn promise that no arms or emigrants would be allowed to pass through its territory on the way to Israel.

There was, however, another side to the Dutch policy; for concession was often granted only when the event was beyond control or when loss of life seemed certain. In October, one of the fedayeen (who would later be freed by the November 22 hijacking) together with three other prisoners seized a chapel in the penitentiary near The Hague. Using smuggled-in arms, they held twenty-two hostages, including members of a visiting choir and some of their relatives. They did release seven hostages but demanded a plane to take them out of Holland along with the other Palestinian then in the prison hospital (who, however, showed no interest in leaving). Finally, after 105 hours, a team of Dutch marines and police stormed the chapel, overpowered the exhausted prisoners, and freed the fifteen captives. The whole operation was quite effective, but both Palestinians were released later in November after the British jet was hijacked.

Then on December 2, 1975, the Dutch were presented with an entirely different dilemma. Seven South Moluccans, descendants of natives of the Spice Islands, who had fled to the Netherlands when Indonesia became independent and crushed their separatist movement, seized a local train in the northern Netherlands near Beilen. They shot and killed the engineer and one passenger and then settled in for a siege. The South Moluccans covered the train's windows with newspaper and informed the Dutch authorities of their demands. The Dutch government, according to Justice Minister Andreas A.M. van Agt, decided almost at once that there would be no concessions and that the terrorists would not be allowed to leave the country. The security forces settled in for a patient siege. In response the South Moluccans killed a hostage. The siege went on. Six hostages escaped through an empty car. A few others were released with messages. Mediators arrived from the forty-thousand-person South Moluccan community.

By then matters had grown considerably more complicated. On December 4, seven other South Moluccans seized the Indonesian consulate in Amsterdam, took twenty-five hostages, and announced a list of demands to the Indonesian government. One Indonesian consulate official jumped from the third-floor window and was fatally injured. The South Moluccans did release four children from the consulate but continued to hold twenty-one hostages. In both cases the Dutch maintained the patient siege, and after twelve days and the onset of a bitter cold spell on the bleak plain north of Amsterdam, the seven terrorists came out of the train and surrendered. Those in the consulate finally gave up after sixteen days. The seven on the train each got fourteen-year prison sentences; those from the consulate received six-year sentences. The general feeling in the Netherlands and elsewhere was that the Dutch strategy of extended negotiation while refusing to compromise had been highly effective.

The techniques and tactics of negotiation learned during the various incidents in the Netherlands and during the Herrema case in Ireland allowed Dutch officials and in particular Dr. Dick Mulder, a psychiatrist and army officer, to establish the basic tenets of hostage-bargaining with revolutionaries, similar in most ways to the practices developed by the New York Police for bargaining with criminals and psychotics. Rather than adopting the no-compromise posture of silence on the part of the Irish and British police, the Dutch felt that negotiations—especially on nonsubstantive matters—would gradually transform the terrorist situation into one where surrender would become inevitable and the hostages' safety certain.

Eighteen months later, Mulder was to have ample opportunity to apply his experience. On May 23, 1977, six heavily armed South Moluccans seized an elementary school at Vovensmilde, four miles southwest of Assen, and took 105 children and 6 teachers hostage. Simultaneously, seven other South Moluccans stopped a train near the village of Onnen, six miles north of Assen, and held fifty hostages. Prime Minister Joop den Uyl went on national television and denounced "a horrible act of terror" and urged self-restraint.

Mulder and his associates moved in to begin the patient siege. Minister of Justice Andreas van Agt announced on television that the government would not agree to the demand to release the twenty-four South Moluccan prisoners and fly them to an unspecified country in a Boeing 747. And so the siege dragged on day after day. During the second week some of the hostages were released—and many of the children in the school contracted a mysterious disease. The Moluccans in the school freed the 105 children and one teacher. But they still had four hostages. Dutch officials remained convinced that delay, restraint, and the lessons of the past would prove effective.

Ultimately, mediators satisfactory to the terrorists were found, but their efforts led nowhere. The siege dragged on and the gunmen showed no interest in surrendering. The Dutch government announced that the siege could not continue indefinitely, that there would be no concessions, that the Moluccans must surrender soon. The South Moluccan mediators urged the government to have patience, but after three weeks Dutch patience had run out. At dawn on June 11, Dutch marine commandos backed up by five Lockheed F-104 Starfighters attacked the school and the train simultaneously. The was no trouble at the school—the four hostages were released and the South Moluccans were taken into custody. At the train, the commandos shouted to the hostages to get on the floor and stay there just before they burst into the car. Smashing aside the door, they swept the car with gunfire, killing two hostages who sprang to their feet. The final toll was eight dead (six Moluccans and the two hostages) and twelve wounded (nine hostages, one terrorist, and two marine commandos).

The government's justification for the attack was that the health of the hostages was bad, especially their psychological condition, and that matters might get worse. The South Moluccan mediators were greatly disturbed; they felt that patience would have paid off, and they were not alone in their opinion. Prime Minister Joop den Uyl, however, defended the government action on a national radio broadcast: "The Government could not wait any longer—we had to use force to prevent worse from happening. . . . dead have fallen among the hostages and among the hijackers. . . . ultimately we did not see any other way, and we could not and must not have let the hijackers leave the country unpunished."[1] The Dutch had come full circle, from concession to the Japanese Red Army to repression with no concessions (albeit after three weeks).

Flexibility

Although the Dutch had compromised when they could not avoid it, national policy had come to favor a siege plus the refusal of South Moluccan demands. Other democracies shifted from concession to repression and back again, depending on the analysis of the moment. While Sweden had entered the decade of terror as an advocate of national liberation, it switched to advocacy for the rule of law—including emergency regulations as a means of limiting terrorist operations. From 1970 to 1972, protestors who seized embassies were persuaded to surrender. The two Croatians who assassinated Yugoslavian Ambassador Vladimir Rolovic on April 7, 1971, were tried, convicted, and sentenced to long prison terms. Then came the spectacular of September 1972 when

[1]*New York Times,* 12 June 1977.

the three Croatians hijacked the SAS DC-9 and persuaded the government to release six of their colleagues. The next disaster came with the seizure of the West German embassy in Stockholm in April 1975 by five men and one woman of the Baader-Meinhof Group, who demanded that the German government release the Baader-Meinhof members held in jail. During the ensuing confrontation two hostages were murdered when the government refused to release the prisoners and the embassy was dynamited by the terrorists. One of the terrorists was killed in the blast, and the rest were immediately arrested and expelled. This in effect was a German battle held in Stockholm while the Swedes watched.

In fact Swedish antiterrorist legislation was drafted particularly to monitor the emigré population rather than restrict the liberties of Swedish citizens. On April 1, 1977, the law's effectiveness seemed to have been fully demonstrated when the Swedish police detained thirty suspects who had planned to abduct former Immigration Minister Anna-Grete Leijon and trade her freedom for that of the Baader-Meinhof people in West Germany. On April 3, the two Germans involved—both well-known radicals—were turned over to West Germany, and eventually twelve Swedes were indicted and two others remained under detention. There was general Swedish and German satisfaction but also a price to pay.

In Germany, on April 4, Chief Public Prosecutor Siegfried Buback held a news conference to announce that warrants had been issued against the two extradited men. Coming as it did when the twenty-three-month trial of three leaders of the Baader-Meinhof group (Andreas Baader, Gudrun Ensslin, and Jan-Karl Raspe) was drawing to a close and an inevitable guilty judgment, the arrests seemed to indicate that the day of the urban guerrilla was nearly over in Germany. The leadership was dead or in prison. Holger Meins had died on a hunger strike in November 1974. Ulrike Meinhof had committed suicide in prison in May 1976. Wilfried Bose had been killed at Entebbe. Other members had been captured abroad or disappeared into exile. There were 110 accused or convicted terrorists in German jails. There were only fifty to sixty still wanted by the authorities and now, with Swedish help, there were two less. If the Germans had compromised with Palestinian terrorists, they did not do so with their own.

Whether the hard line "worked"—intimidating the militants—was another matter. At 8:45 A.M. on Thursday, April 7, three days after Buback's news conference, a young couple pulled their Suzuki motorcycle into a service station in the middle of the city of Karlsruhe, not far from the German Constitutional Court. A half hour later Prosecutor Buback's official limousine drove up slowly and stopped at a traffic light near the service station. The couple swung the Suzuki around behind the limousine, idling just back of the car's left rear fender. When the light turned green, the motorcycle's rear-seat rider produced a submachine gun and fired a long burst into the back of the limousine, sweeping the cone of fire through the chauffeur as the Suzuki raced away. The limousine was stalled, the three men inside slumped over, punctured with bullet holes. Thirteen spent cartridges rolled on the street. Buback and his driver were dead. Buback's bodyguard was mortally wounded. The rented Suzuki was found abandoned near the Karlsruhe-Frankfurt autobahn. The West German news agency DPA received a call from the Ulrike-Meinhof Special Action Group taking responsibility: "You'll hear from us again."

West German Chancellor Helmut Schmidt, speaking in the Evangelical Church of Karlsruhe at the memorial service for the three murdered men, asked Germans not to overreact to the murder.

> The murderers want to create a general feeling of official powerlessness. They hope that their violence will bring about an emotionally charged indiscriminate, uncontrolled reaction so that they can denounce our country as a fascist dictatorship. Their expectation will not be fulfilled. Our free way of government could be sacrificed only

> by ourselves. Our moral condemnation of the perpetrators, and anger and shock will not lead us to act out of emotion. In some parts of our society, in some of its institutions and media there was and is an intellectual source of support, fertile ground for ideologies that sanction violence. But terrorists are not misguided reformers—these are criminals, before God and man.[2]

There were those who felt that when no German lives had been at stake during the fedayeen spectaculars the authorities had been most flexible, most unwilling to risk a gun battle or to keep Arab prisoners who might set off a cycle of further hijackings. On the other hand, when the gunmen were natives and their targets high officials, a very hard line had been taken.

Laws had been passed that were, many felt, an overreaction that represented a return to an authoritarian past. In 1976 "terrorist conspiracy" was made a major crime and suspects could be held in jail awaiting trial for up to five years. Courts had the power to seize and read letters between jailed suspects and their lawyers. In the spring of 1977 it was revealed that the authorities in Stuttgart had bugged lawyer-client conversations, suspecting the Baader-Meinhof people and their attorneys of plotting violent crimes. The German authorities responded, as had others, by saying that extraordinary powers were necessary in such an emergency—an emergency heightened by those who sympathized with the expressed political ideals of the prisoners. Foreign Minister Hans-Dietrich Genscher explained that "it should be clear by now that these actors on the terrorist and anarchist scene are not misguided idealists but cold-blooded, vicious murderers."

Even when the terrorists were cold-blooded murderers, as was the case with Baader-Meinhof people in Stockholm (and the South Moluccans in Beilen), the authorities had attempted negotiations, even if there was no intention of concession. Until the Croatian hijacking incident in New York,

[2] Ibid., 14 April 1977.

United States policy had remained no-negotiation and no-concession. And even after the Croatian hijacking, officials insisted that United States policy had not changed.

Then, in March 1977, it became apparent that the new Carter administration had opted for a more flexible approach. On March 9, at 11:00 A.M., six members of the black Hanafi Moslem sect seized a wing of the B'nai B'rith headquarters in downtown Washington, D.C. Three others took over offices in the Municipal Building and three more took control of the Islamic Center. The takeovers were swift and brutal. At the Municipal Building, Maurice Williams, twenty-four, a black radio reporter for station WHUR of Howard University, was shot and killed. City Councilman Marion Barry was shot in the chest. Robert Pierce, fifty-one, a former State Department employee who was studying law, was hit with a shotgun blast. The blast paralyzed him from the waist down and shattered his right arm. At B'nai B'rith, employees were slashed or shot. Even after they had control, several of the Hanafi members brutalized the hostages.

The Hanafis were religious fanatics, led by Hamaas Abdul Khaalis. What Khaalis wanted in essence was an act of vengeance that would give vent to his intense anguish and frustration. Four years previously, on January 18, 1973, seven members of the Hanafi family had been killed; four of his own children had been butchered and dumped in a swimming pool. His wife and daughter had been shot in the head and left for dead. The trauma would have been vast for anyone, no matter how stable; and Khaalis was not stable. He had a long and troubled history of mental instability, and the slaughter in January understandably had a profound effect on him. At the trial of the killers, his outbursts led the judge to fine him 750 dollars for contempt of court. He was—and remained—contemptuous; for, despite the guilty verdict and sentences, he felt that justice had not been done. Now, with 149 hostages, he wanted restitution. The siege was under way.

Actually, the authorities had a wealth of assets. First, they knew who the terrorists were; the slaughter and trial had brought the police in contact with Khaalis and the others. Then, Washington Police Chief Jerry V. Wilson was chairman of a federal task force on terrorism, and Deputy Police Chief Robert L. Rabe, an excellent hostage-negotiator, got in touch with Ambassador Heck at the State Department to exploit all the federal resources. A headquarters was set up, and FBI Director Clarence M. Kelly and Attorney General Griffin B. Bell were drawn into the incident. There was a direct communications link, if need be, into the White House, only a few blocks away from the Municipal Building and B'nai B'rith. The experts began to arrive on the scene: Patrick J. Mullany, the FBI specialist on hostages; two psychiatrists; Robert Blum, a consultant at the State Department; and Steven R. Pieczenik, Deputy Assistant Secretary of State for Management and also a hostage specialist. Thus the effective command-and-control center so often lacking in previous terrorist incidents was available.

Some of Khaalis' demands could be met with relative ease. He wanted his 750-dollar contempt-of-court fine returned. He wanted the film epic *Mohammed, Messenger of God* closed; he claimed it was blasphemous. But his demand that his family's killers be handed over was clearly nonnegotiable and irrational. As the hours dragged on, a surprise initiative came from the Egyptian ambassador, Ashraf Ghorbal, who telephoned Heck and offered his services. Ultimately, after a presidential decision to go ahead, Ambassador Ghorbal, Ambassador Yaghoub Khan of Pakistan, and Ambassador Ardeshir Zahedi of Iran met with Khaalis and convinced him to end the siege. He had his 750 dollars, the film was closed, and he had amply demonstrated that his loss had not been appreciated by others, that real retribution had not been offered. Khaalis had, in fact, during the thirty-eight hours been involved in an intensive session of therapy through violence. When the violence had ebbed, the frustration eased.

As a last concession, Khaalis would be arraigned and then released without bail on his own recognizance.

Superior Court Judge H. Carl Moultrie, following the advice of Attorney General Bell to keep the terms of agreement, did in fact free Khaalis without bail, on his own recognizance. Eventually, he was arrested for threats made over his tapped telephone.

The response to Khaalis' release on March 11 varied. Some felt the government had to keep its word in order to have leverage in subsequent situations. Others felt that any promises extorted during a reign of terror were invalid; still others, that psychotics holding hostages with or without legitimate grievance were really interested in the therapy of violence rather than substantive concessions. The fact remained, however—a proven homicidal maniac had been released as a result of a federal government decision at the highest level to fulfill an extorted promise. Senator Robert C. Byrd of West Virginia was appalled: "It is abhorrent in our society that individuals can commit these atrocious crimes and then be out on their own recognizance." Senator Lloyd Bengsten of Texas echoed Byrd's sentiment. And Fred Herzod, Dean of the John Marshall Law School in Chicago, called the decision to free Khaalis "absolutely appalling."[3] Others insisted that the siege had ended, that lives had been saved, that the government had to keep its word.

Everyone drew comforting lessons that fit their portion of the common wisdom; but the dispute, mainly over tactics, had missed a major point. The United States government had negotiated and made concessions; it had introduced what many had long felt was a needed flexibility in response to provocation. The United States government had not, however, fashioned a firm new position on terrorism. The next time—and most agreed there would be a next time—all decisions would again be made on an ad hoc basis, probably by

[3] Ibid., 13 March 1977.

those who were far from the event, ignorant of the gunmen's identity, reacting on the basis of the common wisdom as supplied by those convenient to hand.

Concession and Accommodation

Obviously many nations, if not many democratic nations, abetted favored revolutionaries, applauded their victories over old foes, offered sanctuary or equipment and always sympathy. Some more discreet, fearful of retaliation or a bad press, turned a blind eye to their enemies' enemies. Obviously, the punishment of Palestinian fedayeen in most Arab countries was cosmetic, at best brief detention and often a quiet hero's welcome. And all too obviously the mills of justice in most democratic countries did not grind very finely when the fedayeen fell into the law's hands.

This was as true in Greece or Germany as in Cyprus, a small state with truly serious internal problems of its own. The Cypriots wanted to be left alone. The Greek government, however, became increasingly disenchanted with Cypriot leader Archbishop Makarios. They saw him as a crypto-communist who opposed the efforts of the EOKA-B organization to merge Cyprus with Greece. The new Greek junta leader, Brigadier Demetrios Ioannides, a narrow, fanatic man who thought only in slogans, decided in July 1974 that Makarios must go. On Monday morning, July 15, Greek army tanks stationed on Cyprus began slamming shells into the presidential palace. At 2:55 that afternoon, with heavy fighting continuing and Makarios still free, Nicos Sampson, an old EOKA gunman, was sworn in as the new President of the Cypriot Republic. The Greek colonels had in fact annexed the island by planting their man Sampson. His "elevation" presented Ankara with a challenge that could not be easily avoided. On July 20, Turkey invaded Cyprus, seizing a large beachhead on the northern shore and linking up with the Turkish quarter of Nicosia. After peace talks had collapsed, the Turkish army advanced further against limited resistance and seized forty percent of the island—including Famagusta—before another cease-fire went into effect. The Greek Cypriot community had been devastated. There were forty thousand Turkish soldiers on the island, thousands upon thousands of Greek refugees, and more thousands missing. Sampson was in hiding, Makarios in exile, and the Turks triumphant.

It was too painful to contemplate. The major culprit, the unsavory Ioannides, had been removed, together with the rest of the colonels. But they had left Cyprus in ruins. That it had been Greeks that provoked disaster was forgotten by many, especially those in EOKA-B who felt they had been betrayed. Probably the Americans knew of the coup and had approved of the colonel's adventure. The CIA certainly must have known. So for EOKA-B it was as much the fault of the United States that Cyprus had been ravaged as that of the Greek junta; more so perhaps. On Monday morning, August 19, a large crowd of anti-American demonstrators appeared in front of the U.S. embassy, screaming, cursing, waving placards, and trying to force the steel gates.

Fifty yards away, on the side of the embassy, several men had scrambled up to the third floor of a vacant building. They moved up to windows overlooking the embassy. They could not see inside, for the embassy's windows had been taped with blackout paper and the drapes were drawn. All of the men opened fire simulatneously with Soviet-made AK-47 automatic assault rifles. The gunfire tore into the corridor window, smashing the glass and hitting Ambassador Rodger P. Davies. He slumped to the floor, dead. Scrambling to escape the burst of fire, Antoinette Varnava, a Turkish Cypriot employee, was also hit and killed. Outside, the gunmen, still unaware of the effectiveness of their fire, rushed from the vacant building and disappeared.

It has hardly a clean escape or an elegant operation. The American Broadcasting Company, which had been covering the demonstration, had the assassins on film. Within hours the American CIA, using old contacts and frightened or sympathetic informers, had an excellent idea of the identity of the killers. The Cypriot government did actually arrest three men suspected of involvement. But, they assured the Americans, those arrested were not EOKA-B gunmen but simply fanatical young men mimicking the methods of revolutionaries. The Americans were unconvinced. As time passed, the Cypriot response seemed lethargic. The police were still heavily infiltrated by EOKA-B members. Evidence disappeared and no new arrests were made. The ABC film seemed of no use to the investigators. Nothing happened until February 1977, over two years later. Then, with the new Carter administration in power and the need for an American initiative to move Turkey, Cypriot authorities charged six men with the killing of Ambassador Davies.

The suspects were well known in Cyprus as members of EOKA-B. The most notorious was Neoptolemos Leftis, a small, brown man of fifty with a vast handlebar moustache, who had been Nico Sampson's bodyguard for a brief time. Leftis had also been behind a spectacular bank robbery conducted with the aid of a tank during an earlier outbreak of violence. Two suspects, Ioannis Ktimatias and Loiza Savva, were policemen. These three suspects were charged with manslaughter; the other three were charged with lesser offenses. Diplomats on the island noted that the charges marked the first time a foreign government had made a determined effort to prosecute persons implicated in the murder of an American diplomat. The Cypriot government might have moved slowly, but at least and at last there had been movement. Unlike similar cases in Khartoum and Beirut, the Cypriots seemed intent on compiling a serious case.

Then the momentum of the exercise began to wane. The number of defendants dwindled. Two were acquitted, one on a technicality. Two others were sentenced to a few months in jail after pleading guilty. That left only two—Leftis and Ktimatias. By the time the case had been reopened, during the summer of 1976, almost all the original evidence had disappeared from police files. Several witnesses reported privately that attempts had been made to intimidate them. The testimony of a key prosecution witness was rejected by the court because it differed from the evidence he had reportedly given the police. The court also refused to allow the prosecution to introduce the ABC news film. There was no mention of EOKA-B membership at the trial. The audience packed into the three rows of public benches was overwhelmingly sympathetic to the defendants. On June 3, Judge Demetrios Demetriades, speaking for the three-judge panel, announced that not enough evidence had been produced in court to link the last two defendants to the incident.

Greek Cypriots were hardly surprised. It would have been a violation of tradition to convict men on a serious and complicated charge for a political offense. The two *were* convicted of two lesser charges. Anyway, the forms of justice had been followed and there had *not* been enough evidence.

On November 20, 1975, after a long illness, Spain's Generalissimo Francisco Franco died and nothing could ever be the same in Spain again. Spain's new leader, King Juan Carlos, had no intention of maintaining the authoritarian, neo-Fascist organic state that he inherited. But movement away from the national state that had dominated Spanish political life for forty years was certain to be resisted by those who benefited from the system, including the security forces and the military. They would see little reason to change; and they already suspected treason. Spain, then, from the death of Franco to the ultimate parliamentary elections in June 1977, was in an era of transition between authoritarianism and democracy. The structures and skills of repression eroded erratically, while the benefits of dissent, free organization, and an open society appeared sporadically and often uncertainly. Spain was assuredly not an effective democratic society without a nationality problem, but neither was it a brutal and effective authoritarian regime.

In July 1976, the King announced an amnesty for all those convicted of nonviolent crimes. On August 4, three members of the illegal Spanish Communist Party and six Basque nationalists were released. By the end of the first week, 41 out of the 630 prisoners covered by the amnesty had been freed. The ultra-Left was unimpressed. The Basques pointed out that most of their prisoners had been convicted for violent acts and were not eligible for the amnesty. The Right felt betrayed. Adolfo Suárez, whom the king had appointed chief minister, proceeded cautiously, dismantling the machinery of repression step by step. Colonel Federico Quintero, the hard-line police chief of Madrid, was eased out. The Social Political Brigade, Spain's political police, was gradually closed down. In an effort to placate the Basques, the government lifted the Franco decree of 1937 that declared the Basque heartland was to be punished for resisting the national movement. The move had no apparent effect. The Civil Guard in the Basque country continued as before with sweeping searches, extensive arrests, and bloody shoot-outs. Suspects were still tortured and Right-wing gunmen tolerated. There were riots, demonstrations, confrontations, and ambushes. And it was not only the Basques. On October 1, an unknown organization had shot and killed four policemen—henceforth they called their organization the First of October Anti-Fascist Resistance (GRAPO).

On December 11, 1976, four days before the plebiscite on new governmental forms, GRAPO kidnapped Antonio María de Oriol y Urquijo, President of the Council of State, former Minister of Justice, a Basque, and one of Spain's wealthiest men. The kidnappers demanded that fifteen political prisoners be released or Oriol y Urquijo would be executed. On

December 17, Interior Minister Rodolfo Martín Villa went on television and broadcast the government's refusal to yield to the kidnappers' demands.

The new year came in with a wave of violence. On January 25, two gunmen in their mid-twenties entered the third-floor office of the Communist Party labor organization, Atocha Workers Commission, one of the many parties and groups that were gradually coming out from the underground. They sprayed the room with submachine gun fire. Two Communist labor lawyers and a secretary were killed. Six other lawyers were gravely wounded, two mortally. Yet, the odd thing was that by January it was relatively clear that the welter of radical organizations, especially on the Left, had no real constituency; they represented no one but themselves. And the Right-wing gunmen, who mainly limited themselves to threats and arson, were again no real threat to the state, although the military might be if the political violence were to escalate.

The real unresolved problem was the matter of Basque separatism. Since 1958, the Basque militants of *Euskadi Ta Askatasun* (Basque Homeland and Freedom), ETA, had waged a violent campaign that had ebbed and flowed in relation to the level of repression, the trauma of internal disputes, the availability of French sanctuary, and contingencies of the moment. After Franco's death in November 1975, the militant ETA leadership saw no reason to give Juan Carlos a period of grace. The ETA operations outraged security officials. Manuel Fraga Iribarne announced "if terrorists want war, they will get it—and with all the consequences."[4] Juan Carlos and his chief minister, Suárez, however, wanted a democratic compromise rather than further repression. For ETA militants, even the December electoral results that revealed the prospect of a democratic Spanish parliamentary system made little difference. The King and Suárez might ac-

[4] Ibid., 14 May 1976.

commodate Basque separatism to some degree (particularly after the June elections) with a mix of promises and concessions and delay; but in the meantime, ETA was uncompromising. And in the Basque country ETA was not simply tolerated but protected. Most Basques might not want to become involved in the struggle, but most had Basque, not Spanish, priorities and most abhorred the Spanish security forces.

The Basque "solution" of the slightly democratic government of Suárez was not exactly novel—what could not be absorbed or co-opted or coerced must be expelled. Although the Basques, even the moderate Basques, insisted that an amnesty should include *all* political prisoners—even those involved in violence—this was difficult for many in the security service to stomach. Still, at least a beginning step was taken toward easing Basque hostility. On February 14, 1977, four Basques were released; but there were still 110 in prison. And on March 8, the Civil Guard in Guipuzcoa Province shot and killed two ETA members after stopping their car at a roadblock. They claimed to have fired in self-defense, but no Basque believed them. Two bishops in San Sebastian expressed doubt about the incident and called for full amnesty. On March 11, the government proclaimed a new amnesty, but in vague terms. Off the record, Foreign Minister Marcelino Oreja Aquirre, a Basque, indicated in April that about eighty of the Basque prisoners would be released. The Basque response was not promising. "We cannot believe a Basque traitor—Oreja. The only thing that we can believe in is the popular struggle."[5] There were continued demonstrations demanding full amnesty. In May there was a general strike that closed down the four Basque provinces. Five demonstrators were killed and fifty-seven people were injured, including twenty-four police officers.

More security forces were sent in; but toward the end of

[5] Ibid., 3 April 1977.

May, the more unpalatable Basque prisoners were released and flown out of Spain into exile. During the last week of May and first week of June, six Basques were released and flown to Belgium and Norway. On June 6, two more guerrillas serving jail sentences were flown in a Spanish Air Force plane to exile in Norway—they were two of the best known of the sixteen remaining Basques serving jail sentences. The assumption was that sooner rather than later the last Basque would be out of Spanish jails.

The Spanish response to the Basque separatist problem was hampered more than anything else by the chaos and confusion within the government during the period of transition. Suárez's government had to establish a legitimacy beyond the King's word, placate the more suspicious on the Right, repress the Ultras while reforming the security forces—and, of course, run the country, collect the taxes, attract investment and appear at appropriate ceremonies. To buy time and peace in the Basque country during the electorial campaign, Suárez, who had entered the elections with his Union of the Democratic Center—that is, the government—at the last moment, avoided the humiliation of granting full amnesty to the men of violence but shipped them out of the country. It seemed a small price to pay.

In the Spanish case the key to the mix may have been the concession of amnesty-and-exile to buy time. Elsewhere, a democratic country may represent a similar response as a necessary sacrifice to protect lives rather than as an accommodation to violence. In the case of Mexico, many of the prisoners extorted out of Latin American jails by kidnappers have flown in to certain sanctuary. Exile is a long-time feature of Latin American politics and sanctuary a tradition. Few see any parallel between welcoming a hijacker and allowing asylum to freed political prisoners of congenial ideology. The latter appear as victims while the former are seen as villains, unless, of course, their ideology is congenial. Even to suggest that the granting of asylum means complicity in political violence would be resented because by supplying a haven Mexico can simultaneously save American or Japanese or German diplomats and free the victims of authoritarian regimes.

On January 23, 1973, the United States ambassador to Haiti, Clinton E. Knox, was seized by two men and a woman while driving to his residence in the hills outside Port-au-Prince. Taken out of his car at gunpoint, he was driven in another car to the residence. After a telephone call from Ambassador Knox, the United States Consul General Ward L. Christensen arrived at the residence and was also taken hostage. The kidnappers, linked to the National Brigade—"revolutionaries trying to liberate our oppressed people"—demanded the release of thirty-one prisoners, safe passage out of Haiti, and at one point a ransom of 500,000 dollars. The incident seriously embarrassed the unsteady government of twenty-one-year-old President Jean-Claude Duvalier, "Baby Doc," who could not afford to alienate the United States. Still, the Haitians insisted that only twelve on the list were actually in prison and the kidnappers finally agreed to take the twelve. American Secretary of State William P. Rogers flatly refused to consider the ransom demand of 500,000 dollars, whereupon the Haitians, apparently on their own, came up with 70,000 dollars, which the kidnappers accepted. In Mexico City Foreign Minister Emilio O. Rabasa announced that Mexico, after consultations with the United States and Haiti, had agreed to give the kidnappers and the freed prisoners political asylum. Knox and Christensen were released, and three kidnappers and twelve prisoners were flown to Mexico. On arrival they were, indeed, given sanctuary, although the ransom was confiscated by Mexican officials. In Mexico the gesture was seen as humanitarian rather than complicity in terrorism. It was a view one assumes was shared by Knox and Christensen, whose own government would neither negotiate nor concede to blackmail, although it would allow others to do so—in this case, Baby Doc.

Several governments have insisted that for a legitimate

democratic government to concede to some or all of the demands of the men with the guns is a policy of magnanimity quite unlike the arrogance of "no-no-never." A refusal to bargain does not *make* the terrorist murder his hostage, but it greatly enhances the possibility. Thus, to say "no" was to be an accomplice, even if at a great distance. As always there is no easy answer, as the experience of Austria—the country that most readily conceded in the name of humanity—indicates.

The major reason for considering Austria a soft target for terrorists had been the September 1973 incident when two Arabs had extorted a promise from authorities in Vienna to close down the Schönau transit camp for Russian Jews on their way to Israel. The Austrians, on the other hand, felt that their concession had persuaded the Arabs to release the hostages, whose safety was Vienna's primary responsibility. But Austrian jubilation at the release of the hostages soon turned sour. The Israelis were outraged at the concession, and they were not alone. The United States government expressed dissatisfaction, as did others. A week of international scolding followed, ended only by the outbreak of the October War in the Middle East. Austria felt misunderstood and abused.

Two years later the Austrian authorities faced another incident that made the September 1973 operation appear trivial in its international implications. On Sunday, December 21, 1975, a group of young people, one in a Basque beret and an open white trenchcoat, entered the building of the Organization of Petroleum Exporting Countries (OPEC), then meeting in Vienna. The reporters loitering in the lobby noticed that they all carried sports bags and that one was a small young woman in a maxi dress with a gray wool cap over her head. They did not look like OPEC people and they were not. The man in the white coat was Ilich Ramirez Sanchez—Carlos-the-Jackal—the man who had shot the three French policemen in June 1975 and became the media's archetypal terrorist. The woman was German—perhaps Kröcher-Tiedemann, perhaps Mechthild Rogalli. The other German was Hans-Joachim Klein, a member of *Bewegung 2 Juni* (2 June Movement), a rare working-class convert. Two others spoke Arabic with Palestinian accents, and the last could have been German or Latin American. They were the epitome of transnational terrorists, and their target, appropriately, was an international organization. They intended to kidnap the leadership of OPEC, fly off to sanctuary, and demand a vast ransom. In the process they would punish the orthodox and conservative oil producers, Iran and Saudi Arabia, the enemies of their patron, Colonel el-Gadaffi, and their commanders in the PFLP.

Four minutes after Carlos drew a submachine gun out from under his trench coat and asked, "Where's the conference room?" three people were dead—two killed by the girl, and one by Carlos. The German Klein had been shot in the stomach and very seriously wounded; but the terrorists controlled the conference room and the ambassadors—including Sheikh Ahmed Zaki Yamani, all too well known to the West after the Arab oil boycott of 1973 and the incredible increase in oil prices that followed.

Outside, the Austrian Special Command, *Einsatzkommando,* received the alarm a few minutes after the shooting and moved into place. Negotiations opened. Carlos wanted a manifesto broadcast in French on Austrian radio and television every two hours; he also demanded a bus with its windows curtained to be outside at 7:00 A.M. the next day to take them to a fully fueled Austrian Airways DC-9 with a crew of three. If the manifesto were not broadcast, one of the hostages would be shot. Carlos had seventy hostages and an impregnable position. Eventually the manifesto was broadcast, fifty minutes past the deadline. Chancellor Kreisky decided to try to avoid bloodshed. An attack on the building was futile; an attack on the way to the airport or at the plane, far too dangerous. He decided to negotiate the best he could. And he got some results; for Carlos took only forty-two hostages (none of them Austrians), but all the twelve oil ministers and their staffs of secretaries and interpreters. Klein, who had been taken to a hospital in serious condition, would be removed with a volunteer doctor to the plane—"I don't care if he dies on the flight. We came together and we will leave together." At approximately 8:00 A.M., they left together. The red and white Austrian DC-9 took off. Kreisky had arranged with President Houari Boumedienne for Algeria to accept the terrorists and hostages, including Belaid Abdessalam, the Algerian oil minister. In the cockpit Carlos told the Austrian captain ". . . violence is the only language the Western democracies can understand."

Whether the West understood violence, Carlos certainly did. For him, violence had paid. He was on his way first to Algeria and then to Libya, keeping Sheikh Yamani and the Irani Minister Jamshid Amouregar until a huge ransom was paid by the Saudis, Iranis, and the Austrians. The latter transferred a sum estimated from between 5,000,000 and 50,000,000 dollars from a Swiss bank to Aden. Along with the Arab and Iranian payments, the total was probably a world-class record in ransom, surpassing the previous record of the Argentine guerrillas. This time there was no criticism of Kreisky or Austria. What else could have been done?

In the years between Munich and Vienna, the West had indeed come to understand violence; however, this has not produced a consensus on how to respond effectively or even on the nature of the new terrorism. While a variety of techniques and technologies had been devised that could to a degree deter terrorists or ameliorate their violence, there was still uncertainty as to the appropriate tactics or the prospects of a general strategic response. Nearly all the threatened would have *liked* to maintain a "no-no-never" posture, a hard line that would insist terrorism was absolutely illegitimate, counterproductive to the avowed cause, and beyond accommodation. Those who urged if not concession at least a hearing for the desperate gunmen were usually to be found in the rebel camp or in an ivory tower. Even if their advice might be sound, no premier or president liked to have his attention focused on just grievances by the sound of gunfire. Still, "no-no-never" was not always possible—except as far as the Israelis were concerned. Their strategic response was the only truly consistent reaction—one that included the expansion of their military doctrine of retaliation with the principle of no-concessions included, a reaction founded on their perception of the nature of the war with the Arabs. Matters were more complicated elsewhere.

Increasingly the Western governments approached the problem on various levels with considerable areas of agreement. There was a feeling that terrorism would not simply disappear—even if, say, the Palestinian problem could be "solved"—and that at least some terrorists would be beyond accommodation. On the broadest plane, the West sought in a variety of international forums to make such violence illegal and unsavory, more particularly to do so with special agreements that made such terrorist acts punishable. Whatever the apologists of the Third or Fourth World might say, terror in the West was an abomination. Coupled with this there were general tactical efforts to make such violence difficult: security hurdles had been raised, new forms and forces established, air passengers filtered, intelligence improved, and new technologies developed. Yet with each new incident the differences in response between the flexible and the adamant revealed that, in the middle ground between international covenants and air marshals, threatened governments still acted without serious prior planning, their response determined by the posture, principles and politics of the moment. On that level the second round of response was no different from the first.

PART THREE

The Response: Strategies of the Threatened

Die Politik ist Keine
Exakte Wissenschaft.
Otto von Bismarck

It is not only that politics is not an exact science but that the equations also appear to change from day to day, slipping and sliding with imprecision. This is certainly true in the case of the democratic response to terrorism. Even the perception of the threatened may not be static within a single country, and agreement on the necessary tactics of response may not be stable either. Thus, the Dutch, in the case of the South Moluccans, have not decided how to balance strategic accommodation (what will best create a peaceful future for the South Moluccans within the Netherlands) with tactical repression (no concessions to hijackers). And each Western democratic state evolves a response as a result of a mix of factors: habit, history, personalities of the moment, traditional prejudice, intuition, and usually an eye on the public's desires. It is relatively simple, then, to contrast the responses—the United States has the death penalty for air piracy and Sweden a four-year term; France has a separatist problem and Australia none. It is much harder to *compare* the policies of recent years; every case seems special.

The Irish and Italian experience have little more in common than the problem of what to do about gunmen who challenge the authority and legitimacy of the state. The Irish problem stretches back into the past, for the present disturbances have existed in one form or another since the British withdrawal in 1922 from the twenty-six counties that formed the Irish Free State. What has evolved over the years is a traditional means for the center to act upon events. These state acts have been for different purposes and with different results, but the record is long enough and sufficiently detailed so that rather general statements can ultimately be made about the nature of the threat and the response. The Irish mix is, however, very Irish, very difficult to transport, and has very little to do with filtering air passengers or debates in the United Nations. Simply because it is pure, unsullied by the Japanese Red Army or the Palestinian fedayeen or even the trendy ideologies of the moment, Ireland makes a very useful case study of a strategic response. And all "strategic" responses, even the pure Irish one, are more often a mix of tactics of the moment than a grand plan.

The Italian case is much the reverse. Until quite recently, in fact until the spring of 1977, few in Italy fully understood the politics of violence. The identity, intentions, and patrons of the gunmen were in question. The Left assumed a plot by the Right, the Right assumed provocation by the Left. The Center sifted potential guilt for political advantage. Analysis was replaced by slogans: a "CIA maneuver," "Fascist provocateurs," "mad Maoists." There was the assumption that there was political capital to be made from disorder in the streets. And there was increasing disorder in Italy, where every form of political violence, every motivation or ideology could be found: separatists and transnational terrorists and the bombers of the militant ideologies, Left and Right. What was curious about the Italian case was the slow-maturing perception of the threat by those in power, whether in the various party structures, in the bureaucracy, or in parliament. The Irish knew exactly what their problem was. The Italians were divided on the nature of the problem and the need for a response—and they stayed divided.

Thus Irish policy can be covered in sweeping strokes, while in Italy even a chronology of violence sparks controversy. Were the "acts" political? Who was responsible, for what ultimate end, and in complicity with what other forces? When a bomb kills the British Ambassador in Dublin, everyone, including the British, knows that the detonation was political, that Irish Republicans of some variety were responsible, that it was one more "shot" in a two-century war to break the British connection and establish a United Ireland, and that at this stage the assassins probably acted alone, perhaps even in isolation of the IRA, and certainly without the support of a single reputable Irish politician. When an Italian official is killed in Rome, even if pamphlets are scattered about claiming responsibility, no one is sure whether it is an act of Left or Right provocation, a part of a strategy of tension, a deed of vengeance, a part of a plot with ties to the bureaucracy—or even simple murder for personal gain, cloaked as a political event. Consequently, while Ireland presents a pure and simple study of the response to armed subversion, Italy's situation is muddled and uncertain and thus of equal value. Ireland and Italy represent the two ends of Bismarck's inexact science.

CHAPTER

10

THE IRISH EXPERIENCE: DEMOCRACY AND ARMED CONSPIRACY, 1922–1977

It is not those who can inflict the most but those that can suffer the most who will conquer . . .

Terence MacSwiney

All nations are special, their experience difficult to transfer. Ireland often seems more special than most. Despite a heritage of violence and a tradition of rebellion, there has not been for nearly two centuries any real debate on the ideal physiology of government (parliamentary democracy), only debate on the forms of the state (British province, Home Rule, the Republic) and the bounds, thus, of the nation. Since the establishment of the Irish Free State in December 1922, no serious political organization has considered an alternative to democracy. The Irish struggle of national liberation, the first of the twentieth century and archetype for all others, has not only led to a democracy but also to a democracy accepted by nearly all—the fruit of a long tradition, a heritage of compromise and accommodation as well as violence and rebellion.

Yet since 1922 Irish governments have been troubled by the presence of unreconciled Irish Republicans, organized overtly and legally as the Sinn Féin Party and covertly as the Irish Republican Army (IRA)—those dedicated to the use of physical force to break the connection with Britain and establish the *real* Republic. And since 1922 Dublin governments manipulating various strategies of accommodation and repression have sought to achieve as much of the ideal Republic as practical while simultaneously opposing the pretensions of the IRA. Ultimately, the primary obstacle proved to be the existence of the Province of Northern Ireland, organized in the six counties of the northeast as an integral part of the United Kingdom. Two-thirds of the Northern population, military Protestant and loyal to the British connection, abhor the prospects of a United Ireland. Thus Dublin managed in the twenty-six counties to remove the symbols and trappings of the British connection, to cut the Commonwealth ties, but not to end partition. And so with this last grievance, coupled with a general concern for the fate of the Northern Catholic minority, the IRA has maintained both a mission and a means—physical force—that Dublin denied itself.

A great deal that goes on in Northern Ireland, then, has a tremendous impact on Southern events, but can only marginally be shaped by Dublin. After the establishment of the Irish Free State in 1922, there still remained, decade after decade, the dream of the ideal Republic that in public at least all Southern politicians revered. Most of the Irish recognized it was a dream but always some were willing to die for it. An IRA volunteer sought the Republic, denied the validity of the puppet regimes at Leinster House in Dublin and Stormont in Belfast, and sought to break the British connection.

Background

About the only general agreement concerning the present Irish Troubles is that the violence has long roots, at least back to the seventeenth-century planting of large numbers of English and Scottish Protestants in Ulster, sponsored by the British Crown. Their descendants form the loyal base of the Province and the British connection. For many Irishmen the British connection is the source of all Irish ills and must be broken by the only effective means—physical force. These, the "men of violence," heirs to a long revolutionary tradition, are members of the IRA, dedicated to the establishment of a thirty-two-county Republic first declared by the leaders of the Easter Rising in April 1916 but long a nationalist dream. The 1916 Rising failed, the survivors were imprisoned, and most saw the attempt as the last romantic gesture of an Ireland dead and gone. But the separatists between 1918 and 1921 waged the archetypal national liberation struggle—the Tan War. In 1921 Britain, exhausted by the war and frustrated after centuries of the Irish Question, unable to win by acceptable means and unwilling to resort to terror, sought a negotiated settlement. In effect the IRA, unable to win, had refused to lose, thus bombing the British to the bargaining table. What evolved was an Anglo-Irish agreement, accepted by the Irish Dáil (Parliament) sixty-four votes to fifty-seven on January 4, 1922. And from that agreement and what it implied flowed all the subsequent challenges to the Irish state. Its advocates claimed that it gave Ireland the right to be free, its opponents that it betrayed the Republic. According to the dedicated Republican leader Eamon De Valera, the people had no right to do wrong—a nation could not be denied by sixty-four votes to fifty-seven.

The Free State Years

The Anglo-Irish Articles of Agreement tied the new Irish Free State to Britain with a governor-general and an Oath to the Crown for Dáil deputies. The articles also delineated certain British-held military bases and various schedules of payments for land annuities, and incorporated a previous act of parliament, the 1921 Government of Ireland Act, that had established a separate Northern Ireland government in May 1921. But the big issue was the lost Republic, and in its name the IRA fought a brief and bloody civil war with the new Free State Army and lost. The IRA then went underground.

On May 24, 1923, Eamon De Valera sent the volunteers still in the field a final communiqué accepting that the Republic could not be defended by arms so that other means had to be sought. The most apparent means appeared to be the Republican Sinn Féin Party that under De Valera's direction contested seats with the government dominated by the Cumann na nGaedheal. They did so, however, on an abstentionist platform—refusing to take seats in the "puppet" Dáil at Leinster House. The IRA did not disappear but reorganized for another round. In order to counter this potential threat to the Free State, the Dáil in August 1923 passed a Public Safety Act, an extension of existing emergency powers, that permitted the continuation of internment.

This was merely the first step in the confrontation of the Cumann na nGaedheal government with the Republicans. After some initial Sinn Féin successes, it became apparent to the pragmatic that absentionism denied the electorate effective representation and would assure an erosion of electoral strength. Sinn Féin split and De Valera took the pragmatists out to form the new Republican Party of Fianna Fáil on April 12, 1926. Still refusing to take the Oath, Fianna Fáil ran a strong race in the June 1927 general elections, gaining forty-four seats to Cumann na nGaedheal's forty-seven. De Valera—if he and his allies went into the Dáil—might even form a government and win by the ballot what the IRA had failed to do with the bullet. Yet De Valera would not swallow the Oath. And the President, William T. Cosgrave, would make no concessions: if Fianna Fáil wanted to stay out of the Dáil, then they, like Sinn Féin, would become irrelevant. On

July 10, the Cumman na nGaedheal Minister of Justice, Kevin O'Higgins, was assassinated. In the extremely tense atmosphere that ensued, De Valera and Fianna Fáil appeared at the Dáil and signed their names while insisting that they had not taken an Oath. In any case Fianna Fáil as "a slightly constitutional party" was in the Dáil, where their deputies glowered in opposition. A new Public Safety Act was passed and, outside parliamentary politics, abstentionist Sinn Féin soon decayed to a faithful few. But the underground IRA still created a very real security problem for the Free State government. The Special Branch detectives, using the new powers, sought to crush the IRA with some success. Yet large segments of the public if unwilling to sacrifice for the Republic still sympathized with the aspirations of the IRA. In the nearly five years of the Cumann na nGaedheal government, the IRA remained largely a potential rather than a present threat to launch an armed revolt. The Republicans talked about a second round but really did not want one.

The potential threat of the IRA engendered further legislation—in May 1929 a Juries Protection Bill was introduced in the Dáil. For a variety of reasons—IRA coercion being one—juries had shown a reluctance to convict accused Republicans. It was not, however, until the impact of the world depression on an already depressed Ireland inspired radicals to organize that the Free State government felt it necessary to seek additional legislation. The IRA had in the meantime shot several active policemen, intimidated juries, and become deeply involved in various militant organizations dedicated to socialist revolution or communism or worse. Some in the Cumann na nGaedheal truly felt threatened and some members, not unmindful of the approaching general elections, could see political capital in a "Red scare." The result was Article 2A (a Public Safety Bill) inserted in the Constitution in October 1931. This permitted outlawing subversive groups and parties, the arrest of radicals, sweeping searches, and the establishment of Military Tribunals to try the suspects.

The Cumann na nGaedheal government saw the problem of rebellion after the end of the civil war as one of legitimacy. Their opponents must be persuaded or forced to accept the reality of the Irish Free State. If they did not, they faced arrest and imprisonment. Thus the greatest triumph of the Free State was the entry of the "slightly constitutional" Fianna Fáil Party into the Dáil and into the system. Although Cumann na nGaedheal in 1929–1932 chose to see the IRA and the radicals as a threat sufficient to require emergency legislation, any real danger to the institutions of the state from unrepentant Republicans had eroded. The second greatest accomplishment was the transfer of the power to their Republican enemies after the election of 1932. This was hardly remarked on at the time and yet, given the bitterness of feeling and the possession of the assets of the state, Cumann na nGaedheal's smooth shift to the opposition benches remains admirable.

The Éire/Ireland of Fianna Fáil

During the autumn of 1931, the IRA did not react to the arrests by recourse to the gun, in large part because of the impending general elections in February 1932 that gave Fianna Fáil seventy-two seats to Cumann na nGaedheal's fifty-seven. The interned were released. The banned publications reappeared. Article 2A was a dead issue along with the "Red scare."

What followed was a carefully wrought De Valera strategy to steal the IRA assets, thereby ending the necessity for an alternative "secret" army within the state. The Special Branch of the police, an IRA *bête noire,* was reorganized into the S Branch under Colonel Eamonn Broy and staffed with ex-IRA people. A volunteer militia was set up giving potential IRA volunteers a uniform and a bit of money. Compensation was given for civil war losses and pensions for the Republican wounded. A Military Service Pensions Bill was enacted. Old grievances were thus transformed into new loyalties to the government. The government permitted the IRA to harass the Irish neo-Fascists, so the Army Council thus had a "mission."

De Valera dismantled as much of the Treaty settlement as possible. The governor-general was ignored and ultimately replaced by an obscure Fianna Fáil politician. The land annuities were no longer paid. The Oath, of course, was discarded. When Edward VIII abdicated, De Valera swiftly removed the Crown from the constitution and in 1937 a new Irish constitution was introduced, creating Éire/Ireland claiming domination over all thirty-two counties. If it was not the Republic, at least it was not the Free State.

It was increasingly clear to the IRA Army Council that De Valera was not going to opt for the real Republic and was not going to end partition. And it was apparent to the Fianna Fáil government that the core of the IRA could not be co-opted and would not be satisfied with concessions. In 1936 the IRA Chief of Staff was arrested under Article 2A, no longer a dead issue. The following month the IRA was declared an unlawful association. Thereafter, nothing went right for the IRA. There were damaging arrests in the North. In July the Spanish Civil War began and many Irish Republicans left to fight for the Spanish Republic. The movement remained split between the Left radicals and the Center military types. There was constant seepage into Fianna Fáil or retirement. In 1938 an Anglo-Irish Agreement returned control of the British military bases to the Dublin government—a tremendous victory for De Valera and, many thought, a harbinger of an agreement to end partition. Seemingly the IRA had become, like Sinn Féin, irrelevant.

In fact the IRA entered a period of new militancy, presenting unexpected and unwelcome dangers to the state. On January 12, 1939, the IRA Army Council sent the British an ultimatum to withdraw from Ireland and three days later, when there was no response, issued a Proclamation of War. The major arena of IRA activities during 1939 and 1940 was Britain. IRA active service units detonated a long series of sabotage bombs, usually small, seldom effective, and without any real strategic purpose. British prisons began to fill up with those involved in "terrorist outrages," and except for the odd vicious incident—the worst being an explosion in Coventry that killed five and injured sixty in August 1940—life went on as usual. The major public concern was the war in Europe that began in September 1939.

In Dublin, in December 1940, the IRA under the direction of Stephen Hayes raided an arsenal in Phoenix Park and cleared out most of the "other" Irish army's ammunition. But what concerned the De Valera government was not so much this raid but evidence that the IRA had close contacts with Germany—Britain's foe and therefore a potential Irish ally. De Valera was determined on neutrality, a posture that would at last indicate complete separation from the United Kingdom. Any IRA-German connection could easily prompt a British invasion. Thus the IRA seemed to De Valera to be dangerous and violent, waging a bombing campaign in Britain, dragging in German agents, indulging in bank robberies and gunfights.

The government first set up Military Tribunals in 1939 to try IRA suspects. The government at that time found the existing emergency legislation insufficient. The Offenses Against the State Act that permitted the suspension of civil and personal liberties was enacted. Various loopholes were filled when it was amended in 1940. The act largely paralleled the Northern Ireland Special Powers Act of 1922, allowing, for example, the government to imprison a person without holding a trial or making a charge. In 1940 hundreds of known Republicans were interned at the Curragh Military Camp; others were tried and convicted of various offenses. The Special Branch police concentrated on tracking down and arresting those—many old comrades—on the run. To complicate matters for the IRA, members of the Northern Command in Belfast came South, found the organization in disarray, arrested, tried, and condemned Chief of Staff Stephen Hayes. He managed to escape from his former colleagues by leaping out a window and running to the nearest police station. Further arrests followed. There were one or two more gunfights. With tight censorship, IRA activities attracted little notice. The men in prisons and internment camps were largely forgotten. In 1944 the last Chief of Staff, Charlie Kerins, was captured, tried for the murder of a policeman, convicted, and hanged in December. By 1945 the IRA had apparently been destroyed as much by its own splits and schisms and futile campaigns of violence as by the repression of the governments in Dublin, London, and Belfast.

At first Fianna Fáil did not necessarily see a conflict of interest between the IRA and the Fianna Fáil government. As fast as possible, the hateful Anglo-Irish agreement was to be dismantled and as realistically as possible the country moved towards the Republic. In the process, of course, De Valera quite consciously co-opted potential opponents by eliminating old grievances, rewarding old services, opening new doors. While Cumann na nGaedheal had felt there could not be any legitimacy in Republican pretensions, De Valera believed that there could no longer be any purpose in them. When the hard core of the IRA persisted, Fianna Fáil, beginning in 1936, used the same tools of repression as had the Free State. After 1939 when the IRA threat not only to neutrality but also to internal order became serious, Fianna Fáil opted for more stringent legislation, more rigorous repression, and at last an end to sympathy for the misguided—IRA leaders were hanged, imprisoned for extensive periods under primitive conditions, and allowed to die on hunger strikes. Undertaken during a period of strict wartime censorship and in the name of the popular neutrality, Fianna Fáil's policy shift did not result in a similar political shift. The party still sought an end to partition, thus condemning not the IRA but only the IRA's means.

The Northern Campaign

Soon after the end of World War II, two initiatives were undertaken in Ireland to secure the ultimate Republic. With the aid of nationalists in Northern Ireland, Fianna Fáil mounted an international antipartition campaign, employing any available forum and relying mainly on publicity. Another approach was the formation of the Clann na Poblachta Party by Seán MacBride, a former IRA Chief of Staff, and a group of militant Republicans and radical reformers. In an Ireland that was stagnant without economic prospects, after sixteen years of De Valera, there was a tide toward change. In the February 1948 general elections, the Clann won ten seats and came close in others. Then, to everyone's amazement, MacBride formed a Dáil alliance with Fine Gael, the descendant of the Free State Cumann na nGaedheal, that put an end to the Fianna Fáil reign. Despite the presence of the Clann in the cabinet, there was little that the government could do to act on partition. In an effort to get the "Republican" issue out of politics, Taoiseach (Premier) John A. Costello simply declared Ireland a Republic and the Dáil in 1948 passed The Republic of Ireland Act. The British Parliament responded with the Government of Ireland Act that assured there would be no change in the status of the province of Northern Ireland without the consent of the population of the six counties. There the matter rested. Antipartition had fizzled out. The "Republic" had been declared. And in the general elections in 1951 Fianna Fáil came back into office and the momentum of the Clann disappeared.

In June 1954, old Fianna Fáil was again replaced by another interparty government. By then, for the first time in a decade, there was evidence that the IRA had been buried too swiftly. In the North there were two spectacular arms raids on British army barracks in Armagh and Omagh in 1954 and then in December 1956 a guerrilla campaign opened with a series of cross-border attacks. This Northern campaign lasted until February 1962, a low-intensity affair. In the North the Stormont government had depended on the Royal Ulster Constabulary (RUC) and the mobilization of the B-Specials. Suspected IRA people had been interned or tried and sentenced to prison terms. In the South Fianna Fáil, back in office in 1957, had interned suspected IRA volunteers and, when the campaign seemed likely to drag on in 1961, reintroduced military tribunals and long sentences.

During the period 1956–1962 the response of the Dublin government to the IRA had become almost tradition. Arrests, internments, police harassment, censorship where possible, refusal to concede to hunger strikers, and a continuing condemnation of violence. While there was some evidence of sympathy for the IRA, by 1962 physical force seemed outmoded by events. There had not been any great support from the Northern minority. Perhaps the old Republican "issues" no longer had an appeal. Tensions had so far eased that the Unionist Premier Terence O'Neill and the new Fianna Fáil Taoiseach Seán Lemass could exchange visits. Gradualism, a community of economic interest, and the easing of old grievances, bridge building appeared to be the new direction. In 1968 a small but vocal civil rights movement was created seeking Stormont reform, not revolution. Even the IRA had put the gun on the shelf and gone into radical politics.

In the postwar years the only differences in the response of the interparty and the Fianna Fáil governments to IRA provocation was that the latter proved firmer. Neither could make any substantive contribution on the issue—partition—but then neither really felt that the IRA campaign was a very great threat. One or two of the highly emotional demonstrations certainly worried Leinster House; but the old tools of internment, imprisonment, military tribunals, police raids, and the rest did not need reinforcement.

The Northern Troubles, 1968–1977

For the Irish Republic the escalating turmoil in the North after 1968 produced a series of unexpected traumas, kindled old fires, and ultimately a dedication to new realities. The major problem for the Irish government at Leinster House was an inability to act legitimately on Northern events and an unwillingness to stand idly by, mute and futile. The civil disobedience campaign was greeted with complete enthusiasm by Dublin—"our people" in the North had found a means to remove the inequities of the Stormont State. As the shrewd had expected and a few intended, the campaign engendered a violent Protestant backlash by militants who already suspected the moderate O'Neill was going to sell them out to the "Papists." Dublin deplored the violence and urged reform on Stormont and London. London, with the Irish Question back after a lapse of fifty years, responded lethargically. The rising aspirations of the minority made it impossible for Stormont to keep pace with new reforms while damping Protestant anguish. The ultra-loyalists knew that "civil rights" was a code word for United Ireland and "civil disobedience" the harbinger of an IRA campaign. Ultimately in Derry, in August 1969, Stormont could no longer maintain order and the British army had to be moved in to protect the Catholic population of Derry and Belfast.

The response in the South was an anguished concern about the fate of the Northern minority, open to Protestant pogroms, "protected" by Protestant police and a British army that would hardly be congenial to minority concerns. There was little that could be done. It was difficult enough for a member of Fianna Fáil to welcome the presence of British troops in Derry and Belfast. It was quite impossible to send in Irish troops, even if there had been enough. Calls for United Nations action or contacts with Northern Irish political leaders hardly produced the stuff of drama.

What some members of the Fianna Fáil government undertook was to underwrite a dissident segment of the IRA disenchanted with the Dublin Army Council's "political" approach and by their inability to aid the Northern minority in August. By the end of the year, the new Provisional IRA (Provos) had evolved and until April 1970 operated in a tacit alliance with the Fianna Fáil ministers who donated money and endeavored to arrange arms shipments. Those in Fianna

Fáil involved with the Provos were revealed as plotters, indicted, brought to trial in the autumn of 1970, and, as expected, acquitted. By then the Provos, with the Official IRA at their heels, were moving out of the role of Catholic defenders into that of urban guerrillas.

During 1970 in the North, there was a gradual slide into chaos. The security policies of the British Army eroded Catholic goodwill. Sectarian clashes continued. Moderate politicians disappeared. In April 1969, O'Neill had given way to his cousin Chichester-Clark, who had to balance the necessity for reforms urged by London with the need to placate the Unionist militants. The latter wanted the British Army to step in and crush the IRA. The IRA wanted them to try, thus solidifying their position as Catholic Defenders. Matters slipped out of hand. On February 6, 1971, the first British soldier was killed by an IRA sniper. The IRA bombing began: thirty-seven explosions in April, forty-seven in May, fifty in June. Belfast and Derry became war zones. Recruiting to the Provos soared. Desperate, the Stormont establishment urged internment on the British government, not so much to close down the IRA but rather to produce a symbolic "victory" over the rebellious minority. The London government that for so long had done little suddenly decided to authorize internment. The August 1971 sweep by the British Army missed practically all the IRA active service people but collected hundreds of Catholics, including civil rights agitators, but no Protestant extremists. The minority response to the internment "humiliation" was intense. There was a rent and tax strike. Most of the Catholic ghettos in Belfast and Derry became "no-go" zones. The province collapsed into open guerrilla war. There were cross-border gun fights, ambushes in the countryside, car bombs in the cities. More British troops were sent in and yet the violence escalated.

The two years following the arrival of the British Army in August 1969 had been most trying in Dublin. The apparent ministerial involvement with the Provos and the arms smuggling had been scandalous, but understandable. Some who were involved had no apologies. Many in the country condoned their motives if not their methods. In the South the government, thus, felt any moves against the IRA must be undertaken cautiously. The Special Branch did keep a close watch; but with two IRAs, hundreds of new volunteers, and all the unknown Northerners drifting back and forth, they were out of their depth. For many the IRA were the new heroes. Dublin was often jammed with international press and media people. Former Prime Minister Harold Wilson even came to Dublin to talk to the Provos, using a meeting with Irish cabinet ministers as a cover. No one abroad was interested in the real government of Leinster House. For Jack Lynch and the cabinet, the problem was to dampen the potentially explosive emotionalism that might lead to action—any action, no matter how self-destructive. All the news from the North made this difficult: the IRA campaign, the idiocy of internment, the revelation of British Army and RUC torture techniques—and then in January 1972, Bloody Sunday, when British paratroopers killed thirteen civilians during a Derry civil rights rally. Two nights later, while the police stood idly by, a Dublin mob burned down the British embassy.

For Fianna Fáil and subsequent governments, Bloody Sunday was the most dangerous moment. The mob was in the street. The IRA was immune. And the cabinet literally had no policy and no possibility of effecting Northern events. The IRA guerrilla campaign escalated. In March the British government promulgated the Provincial government. The Provos claimed that they had bombed down Stormont. For Dublin the hope was that this meant the time for violence had passed. This was particularly true when on July 7, 1972, the Provos met in London with British ministers and agreed to a cease-fire. The officials had already unilaterally announced their own cease-fire. Although the Provo-British agreement collapsed two days later, Dublin insisted that the time for politics had arrived. Cautious moves began to be made to restrict the IRA freedom of action in the South. The army and police picked up those involved in cross-border attacks. Selected Republican leaders were arrested and questioned. All sensible people in London and Dublin and Ulster urged political initiatives, all-party talks, accommodation.

The IRA kept on bombing. On Friday, July 21, twenty-two bombs exploded in Belfast, killing nine people and injuring about one hundred and thirty. There was general revulsion. In the wake of Bloody Friday, on July 31, the British moved into the Derry no-go zones. Although it was not apparent at the time, the Northern Troubles had entered a new period. First, the IRA could now only pursue a campaign of attrition; the struggle could not be maintained at the July level, but could not be wound down by the British. Second, it was reluctantly accepted that the new Protestant paramilitary organizations had undertaken a campaign of random assassination. The British strategy was to attempt to hold down the violence while fashioning various institutions, proposals, referendums, and initiatives that would permit conventional politics at the center. The Irish response was to encourage all sorts of efforts while hedging about the IRA activities.

Yet every maneuver to create a center for politicians in the North failed. Between March 1973 and May 1975, the North went to the polls seven times and remained polarized. The IRA wanted the British out and a united Ireland. The Protestant paramilitaries wanted no united Ireland, no powersharing, and no compromise, so they continued to commit random murder. The moderate Catholic and Protestant politicians often found their communities alienated, their gunmen with a veto, and their friends in Dublin and London unable to help. Britain could keep the army in place to prevent open civil war, but it could not pacify the provinces nor operate by any other means than direct rule from London. Dublin's only formal involvement had come after Anglo-Irish talks led to an agreement that foresaw some vague Council of Ireland—one of the reasons the militant Protestants refused to cooperate in powersharing. With no Council of Ireland, Leinster House was left urging moderations in the North and trimming IRA wings in the South.

The end of Stormont and the increasing discussions by Britain of an Irish dimension to the Northern problem at least and at last gave Dublin some leverage. Spokesmen claimed that the IRA campaign was no longer needed, that it alienated decent Protestants and inspired the paramilitaries to sectarian murder. The continued IRA bombing and the rising toll of civilian casualties also had convinced many that the Dublin position had much to offer. Many insisted that the Provos should have quit while they were ahead. Various voices were raised in the cause of moderation. And Fianna Fáil decided to act, to take the risk of closing down the Provos. On November 19, 1972, Irish security forces arrested Seán MacStíofáin, Chief of Staff of the Provisionals, under section 30 of the Offenses Against the State Act. He went on a hunger and thirst strike. On November 25, he was sentenced to six months imprisonment. An IRA attempt to free him failed and he was moved to the Curragh Camp. On November 27, the government announced details of a new antiterrorist bill to be introduced in the Dáil. There were further arrests of Republicans. MacStíofáin ended his thirst strike. There were pro-IRA demonstrations, but the government pushed on. Then, during debate on the second stage of the Offenses Against the State (Amendment) Bill in the Dáil, bombs exploded in Dublin killing two people and injuring eighty-three. Fine Gael opposition collapsed and the bill passed by a vote of seventy to twenty-three. In December the President of the Provisional Sinn Féin, Rory O'Brady, was arrested. Sinn Féin offices on Kevin Street were closed. IRA leaders from the North were arrested. The pattern had now been set—the Provos would not be tolerated. In Ireland the Curragh Camp began to fill up. On February 5, Taoiseach Jack Lynch dissolved the Dáil and went to the country—as a law and order candidate.

On February 28, the electorate transformed Lynch into the

leader of the opposition. The Fine Gael-Labour coalition now had seventy-three seats to Fianna Fáil's sixty-nine. As far as repressing the IRA was concerned, the coalition intended to follow Fianna Fáil's lead. Every effort was made to cooperate with British initiatives in the North. Dublin was as frightened of a doomsday war in Northern Ireland as was London, and also as little interested in unity, as concerned with shoring up the Belfast center, and most of all as dedicated to ending the Troubles.

The posture over the next three years simply made Britain's impossible job easier but no more effective. All the elections, proposals, commissions, new groupings, and gatherings of old faces ended futilely. A power-sharing Assembly coalesced in January 1974, and four months later collapsed as a result of a Protestant general strike. In 1975 a new constitutional convention met, wrangled, and ended without a constitution. The problem was there was no solution. The Provos soldiered on. The Protestant paramilitaries continued the random murders.

The Troubles went on and on. There were no-warning bombs planted in British cities. There were book bombs and letter bombs. In the North there were car bombs, some driven by the Provos and some by the Protestant paramilitaries. There were bombs placed in Dublin streets and Irish hotels. And in August 1976, outside Dublin a massive bomb in a culvert detonated and killed the new British Ambassador Christopher Ewart-Biggs. The coalition government seized the opportunity to postulate an Irish emergency and push through further antiterrorist legislation that would increase sentences and narrow the opportunities for subversive publicity. Again civil liberties would be hedged about more narrowly than many thought proper. The most unexpected development was the decision of the Irish President, normally a figurehead, to send the bill before the Supreme Court. The court, albeit reluctantly, agreed to the act's constitutionality. A coalition minister sharply criticized the President's action.

The President then resigned in protest, forcing a presidential election that the coalition refused to contest, anticipating defeat. All the maneuvering ended with emergency legislation that many felt was unnecessary and probably ineffective. The coalition's efforts over three years had been little different from those of previous challenged governments: arrests and sentences (on the evidence of a police superintendent's statement that the suspect had been a member of the IRA); harassment of known IRA members; increased cooperation with British and Northern security forces; increased responsiveness to all moderate initiatives; condemnation of the men of violence and their pretensions. And at the end of the delicate waiting game, they kept faith in the realism of the Irish public and the eventual end of violence in the North. Without the carrot of unification, without even a desire to absorb the troublesome North, the coalition stressed the dangers of unification now, the potential horrors if the Provos continued their wicked ways, and the need for sensible compromises by men of good will.

When the Northern Troubles began in 1968, Dublin faced a problem. In public the state's ideals were the same as those of the IRA. In public and in private, there was universal concern about the safety of the Northern minority. In private there was a fear that the IRA might exploit that fear and the Dublin government's impotence to disrupt the democratic system. Thus the IRA campaign in the North did not challenge the state as had the embittered IRA in1923 or the IRA-German connection in 1940, nor could it be simply closed down by coercion as it had between 1957 and 1962. The IRA after 1969 had a mission that Dublin could not undertake—as Catholic Defender.

So far the Dublin solution has been to use every opportunity arising from shifts in Northern events—and consequently Southern opinion—to limit IRA activities. Even the ministerial involvement with the Provos in 1969–1970 can be seen as an effort by some to co-opt a potentially dangerous organization. The major direction of both Fianna Fáil and, after 1973, the coalition has been to make matters difficult for the IRA while choreographing a change in public attitudes: violence does not pay and is counterproductive; the IRA campaign has set back unification; it is impossible to bomb the million Protestants in Ireland; the gunmen are wicked, cruel, pretentious, and will assure that Northern violence bleeds into the South. There has indeed been a change in Southern attitudes. Many in public doubt the wisdom of a united Ireland. Although the defeat of the coalition by Fianna Fáil in the June 1977 elections, in which their spokesman of the North, Conor Cruise O'Brien, lost his seat, may give some pause. Still, few see physical force as an effective or morally acceptable means. Thus it is not so much the stringent new emergency laws, continued police harassment, threats of censorship, or cooperation with the British government that have changed in the government's response to provocation—but rather in the public political reasoning behind the necessity for such acts.

Review: The Nature of the Response

Any governmental reaction to a lethal challenge almost inevitably includes both "carrots" and "sticks." In the Irish case for fifty years by far the most important was the mix of carrots. By the 1960s, however, all the carrots had been used but one, and that was owned by the British. The major difficulty for Dublin is that this focus of discontent—Northern Ireland—lies beyond the legitimate authority of the state. Except briefly, when the British introduced the Council of Ireland as the Irish dimension, formal Dublin involvement in Northern matters did not exist. More troublesome, until very recently Dublin could not persuade the British authorities in

London of the communality of interests. London assumed, when time could be spared for Irish matters, that Dublin on the subject of unity meant what various officials had said, what the 1937 Constitution claimed. Dublin thus had little leverage in London or Belfast. The Dublin dilemma was not unique but was certainly an extreme example of the difficulties of devising an effective strategy to contain those beyond the reach of coercion or reward. Ultimately, a high degree of tactical cooperation with the British evolved and an often-unstated acceptance of British intentions and strategy. In time Dublin ran out of ideas, clung to the hope that time and exhaustion would permit an effective Northern political initiative, and concentrated on closing down the IRA in the twenty-six counties, where effective tactical repression was possible.

The various Southern "sticks" of repression have been used to effect for sixty years. The present basis for their use is the Offenses Against the State Act of 1939, variously amended and elaborated by subsequent legislation. In essence it all but puts members of the IRA beyond the protection of the law—they can be arrested without warrant, imprisoned without charge or trial, and their homes, offices, organizations, and property become vulnerable to the security forces. On the word of a police superintendent alone they can be sentenced to prison. The latest legislation, passed by the National Coalition after the assassination of the British ambassador, among other features greatly extends the government's privileges of censorship. One of the primary reasons that such gross invasions of civil liberties have been tolerated by the general public is that as long as coercion is directed at the known Republicans, the restrictions are not seen as relevant to Irish society in general, nor unfair for use against those dedicated to dismantling the state. Sympathy for the IRA may or may not exist, but few doubt that those interned or sentenced are guilty as charged, and the state not unreasonable, for usually the state has not acted until IRA activities appear a real danger to cherished goals.

Once the IRA volunteer is imprisoned or interned, his life remains dedicated to the cause. Prison is an opportunity, not a penalty. Efforts by the authorities to prevent IRA people from having special treatment, to punish when possible, and to withold privileges, educational benefits, or visits often tend to harden resistance, producing martyrs. It is, indeed, a policy of the state, quite unstated, to make life miserable everywhere for Republicans. Parents and employers are warned of the dangers of Sinn Féin membership. Suspects are questioned where and when the major amount of embarrassment will occur. Cars are stopped, tickets given, homes searched, members followed for little other purpose than to cause trouble. When arrests and imprisonments are the policy, often those most vulnerable, men with a new job or a sick wife or a heavy mortgage, are lifted first. Once in prison and in the power of the state, the process becomes more intense. Irish prisons—cold, stone, Victorian monsters—are unpleasant at the best of times, with little heat, few recreational resources, limited exercise space, inedible food, and cursory medical attention. At times in response to Republican provocation, prisoners have been allowed to remain naked—if they refused criminal prison garb—denied visits and letters, kept in solitary in cells without furnishings, denied medical attention, denied Catholic communion. The authorities, when any public notice is taken, insist that the prisoners have brought on their own troubles or in some way provoked authority.

Those responsible for IRA prisoners—the prison warders or army guards—rarely have any understanding of "political" prisoners and are permitted or encouraged to see *them* as beyond conventional bounds and in any case dedicated to causing unnecessary trouble: fair game. Outside the prisons, this attitude has gradually been institutionalized in the Irish Special Branch, a force made up of plainclothes detectives responsible for political subversion, often to the exclusion of all else. The regular unarmed police are nearly everywhere respected and seldom concerned with political matters. There has, however, been a noticeable erosion in their restraint with the escalation of Republican activity in the South. There have been accusations of brutal beatings, arrests on unfounded suspicion, coercive interrogation, and use of excessive force during the policing of various demonstrations. The Special Branch remains the cutting edge of the security forces. Yet the dialogue between IRA and Special Branch is a curious one, since Republican policy allows no armed action in the South—the shoot-outs in Dublin during the 1940s have never been repeated. This policy has prevented a cycle of vengeance and retaliation but has done little to moderate mutual antipathy.

Far more important than all the normal tactical police and legislative responses to the IRA has been the government's efforts to transform old Irish attitudes. Every effort is made to paint the IRA as a band of wicked murderers, as criminals who have nothing in common with the litany of old Irish patriots and martyrs. The IRA is illegitimate, brutal, without vision, unwittingly dedicated to destroying the very ideals it supposedly cherishes. And every effort is made to see that the IRA is denied any forum to respond. New legislation will deny Republican spokesmen the "right" to be questioned on Irish television, for example, or even to have their views quoted—a limited ploy, since the easily receivable British television networks have not accepted such a policy. Backed by the immensely powerful and immensely conservative Roman Catholic Church, by the men of property, by most newspaper editors, and by nearly all people of substance or reputation, the government's campaign has clearly had an effect. This is, in part—in large part—because of various IRA blunders. And it is in part because the prospect of a united Ireland containing not only a million embittered Protestants but also (far, far worse) a half-million radicalized and armed Catholics from a betrayed North is a frightening one. In this case the "stick" of reality has been used on the opponents of the IRA whose Northern campaign will endanger the existence of various Southern "carrots.'

Summary: The Foundations of the Lethal Dialogue

The IRA Assets

A Traditional Legitimacy: The foundation that permits the operation of the Irish Republican movement has quite literally been centuries in the making and cannot readily be eroded by right reason or by the grating of a new reality. The Republicans have been strengthened by nearly two centuries of uncompromising efforts to break the British connection. They have claimed, and their claims have until recently seldom been denied by those seeking office, that the Republican revolutionary tradition has been largely responsible for what freedom Ireland has gained and that violence works—that all the democratic maneuvers, political agitation, and civil disobedience would have proved ineffective without recourse to physical force. This legacy of patriots has been accepted in Ireland as real history, not as myth or legend. Those who would deny the benefits accrued through an armed struggle have until recently been in a minority. Even today a great many of the Irish read their past as a long confrontation with British power and consequently accept the present IRA as, deserving or no, heirs to a heritage of struggle. Nearly all the effective symbols of the state and almost all the most gripping patriotic myths authorize this legitimacy.

When, at the funeral of a fallen volunteer, the Provos march slowly to the wail of pipes, impassive under black berets and hidden behind dark glasses, awaiting the final volley over the grave of another patriot and martyr, there is a moment of deep emotion that no rite in the Dublin state can equal. No one feels as deeply about an Irish army parade or the inauguration of a president. For over fifty years the elected politicians have in fact employed the Republican rhetoric. Those who did not largely kept their convictions out of sight of the electorate. So for almost the entire history of the state there was no dissent on the "national issue." The efforts of the major parties to defuse the Republican bomb in no way tampered with Republican ideals and in no sense created a parallel rhetoric. The strongest attraction of the Republican movement increasingly became the obvious hypocrisy of the legitimate institutions. Politicians who served in the "official" Republican party of Fianna Fáil were all too willing to jail and intern those who sought to turn oratory into action. Thus a Republican commemoration is real—a real young man is dead, the flag on his coffin is not just a banner, the volley over his grave is not a sordid ritual but a reward and a promise. The state has been left without the rites of legitimacy. And emotive ritual dramas cannot be easily repossessed or new ones easily created.

The Old Attitudes: What the IRA ritual summons forth in public is an attitude toward history and toward existing institutions. For a great many of the Irish, all with long and exacting memories, there has seldom been an intimate connection between justice and law, law and order. The agencies of order were often considered, justly, as imposed, the laws constructed to preserve others' privileges, and the Crown's justice no more than mean self-interest. Forbidden their religion, paupers in a green land, denied education, advancement, or common decency, few could see a communality of interest with the Ascendency landlords, with the agencies of the Crown. Laws were to be evaded or changed, not obeyed. Concession, compromise, changing times, prosperity, or prospects did indeed erode this posture; but, basically, an abiding suspicion remains. Injustice is no longer institutionalized as it was in Northern Ireland; but still and all, the IRA advocates the ideals long proclaimed and to deny them—and with them, the dreams of the past—is difficult. "Political crime" truly exists in Ireland; and the IRA, whether engaged in train robbery or random bombing, can continue to feed on a considerable pool not so much of sympathy but rather of toleration. The informer, no matter how highly motivated by democratic ideals or by loyalty to the state, remains hateful. Thus Ireland, a really most law abiding state, still is populated by those whose dedication to democracy, to the proposition that now law and justice are truly linked, cannot easily betray those whom Dublin calls "men of violence." This by no means indicates that all the Irish support the IRA or that the Irish ocean largely is friendly to the IRA fish, only that the habits of the past coupled with the situation in the North create an ambivalence. The public tolerates the presence of the IRA.

A New Role: It is in fact the Northern events that give the present IRA a legitimacy related neither to the old rituals nor the old attitudes. Once the civil rights movement had engendered a Protestant backlash, very few in the South trusted the policy of the British Army to defend—effectively, or in time, or perhaps at all—the Northern Catholics. The revelation of the arms trial indicated that the Republican Fianna Fáil had opted to stand back. The victory of the coalition brought to the fore those who at last, if haltingly, denied in public the historical aspirations of the nationalists. There was no longer an inalienable right to unite, and increasingly there was no longer any interest in a unity that would cause chaos and undesired change in North and South. That very few in the South would benefit by a united country and many might suffer, that without harsh sacrifices no Dublin government would be capable of acting on Northern events, all might be true. But the desertion of the Northern Catholic population, whether logical or inevitable, caused grave concern and some guilt. Thus *at least* the IRA was there, doing something, holding a thin line. If the IRA were hounded and crushed, what then?

Thus, the IRA has three factors to exploit: possession of living historical rituals that grant the legitimacy of a long tradition, the reluctance of the population to betray those who advocate the old ideals of the people, and a position as Catholic defenders in the North. It would appear that certainly the first two, and perhaps the last, are wasting assets. Yet old ways and old dreams change but slowly. There is a charm about myths that the new realities of the democratic state cannot quickly or easily erode. Fifty years of political cant are not suddenly discarded. There are years of debts still to fall due. New rituals and myths cannot be fashioned at will, new attitudes summoned up by emergency legislation, and old responsibilities discarded in the name of practicality. Ireland has not always been a very practical country and still can be touched by the old dreams made manifest in the old ways.

The Government's Assets

Legitimacy: The response by the Dublin government in the broadest terms has been directed, often quite consciously, against IRA assets, to stress the real rather than the ideal. The prime effort has been to strike at the pretensions of the IRA while stressing the legitimacy of the democratic institutions of the state. It is a position identical to that taken by Cumann na nGaedheal Minister of Justice Kevin O'Higgins in 1923:

> We will not have two governments in this country, and we will not have two armies in this country. If people have a creed to preach, a message to expound, they can go before their fellow-citizen and preach and expound it. But let the appeal be to the mind, to reason rather than to physical fear. They cannot have it both ways. They cannot have the platform and the bomb.[1]

[1] *Dáil Debates*, vol. X, col. 280.

And as long as the IRA continued to advocate the bomb in the North, the Dublin government sought to deny them a platform in the South. Over and over again the government's spokesmen have insisted that there is no place for secret armies in a democratic state, that no political movement can represent a higher authority than the elected Dáil, and finally in effect that the IRA is betraying the state and hence the nation. The legitimacy of the democratic institutions is everywhere accepted—except by the purists in the Republican movement. It is a very strong card and regularly played. And the Republicans are left isolated with limited electoral support in the best of times and few converts to the proposition that the people have no right to do wrong—that is, abandon the dream of a united Ireland, abandon the North, abandon the old ways for new profits. The government's problem is that many do not see a contradiction in supporting democratic institutions and tolerating the IRA. While the legitimacy of the existing institutions is unquestioned, the presence of the IRA is also accepted—if less universally—although not its claims to represent the nation.

IRA Faults: The government has stressed not simply the pretensions of the IRA in contrast to the reality of the state but also the failings of the men of violence. The leaders are scorned as incompetent, narrow minded, brutal gunmen. Occasionally, there is some indication that a few volunteers might be misled idealists, but for the most part the repeated charges have been that most are callous murderers of the innocent. Dublin accuses the IRA of setting back any hope of unity through the long and vicious bombing campaign that has alienated the Protestants and caused immense Catholic suffering. Violence does not pay and the little cruel men of the IRA have once again disgraced the nation. And there is little doubt that the war in the North is brutal and often degrading, that many accept the government's version of events, that the long tale of horror has left few in the South who believe that the IRA is still embarked on an unstained crusade. IRA counterpropaganda points out that all wars are brutal, but it has not convinced a great many. One horrid incident and the government's case is made again. And more so, since the war has oozed south. Dublin is in danger of violence because of the gunmen; that those gunmen are killers unworthy of Ireland is a theme with increasing takers.

New Pragmatism: Finally, and at last, the coalition government discarded at some cost the patriotic rhetoric of the past. Ministers have indicated that now a united Ireland would be an error. This does not mean that the Irish government has ended advice to Northern Catholic politicians; or given up the concept of an "Irish dimension" to the Northern problem; or even foresworn some ultimate union. Rather, for the foreseeable future, the coalition contemplated the continuation of partition, urging only sharing of power, reforms within Ulster, and an end to violence. A union achieved by violence would be a disaster. Violence is and always has been unnecessary. Some insist that Irish independence would have come through concession, that the Easter Rebellion of 1916 was unnecessary, and the Tan War unneeded. This, so far, has not converted generations raised on the glory days, or Fianna Fáil (swept into office in June 1977), to similar disavowals, but the recourse to pragmatic analysis has quietly convinced many that unification would cost a price not worth paying. Clearly, partition has advantages for the Roman Catholic Church—people's protector in the North and government confidant in the South. Clearly too, no Southern political party could expect to gain much nor would the existing governmental institutions benefit. Thus the coalition government's effort to stress the real world over the mythical was a pragmatism desperately trying to erode the old loyalties before too late.

In a sense, then, the government, especially since 1972, stressed the new realities. They had reason to believe that *their* assets were accumulating, not wasting away; that, as time passes, the old symbols and rituals would become barren, the old loyalties erode, the new gifts within the power of the state attract. The government's defeat, however, may slow this trend, although it is unlikely that Fianna Fáil's response to the IRA will differ greatly—the same familiar sticks are ready at hand. Any vote today would go against the aspirations of the IRA; overt sympathy for the Northern campaign is slight. The democratic institutions engender deep loyalty, the incompetence and arrogance of the IRA (in North and South) is accepted, the dangers of unification now realized. Yet, no one in power rests easy, least of all the new Fianna Fáil government of Jack Lynch—supposedly the *real* Republicans.

Ireland is an effective democratic state with a nationality problem beyond resolution. Britain must "solve" the Ulster problem, not Ireland, and solve it in such a manner that the IRA becomes irrelevant and physical force no longer legitimate in the eyes of the militants. Unlike the aspirations of the South Moluccans, the goal of the IRA—to break the British connection—is feasible. Unlike the ideological goals of the Baader-Meinhof Group or the *Brigate Rosse,* the forms of the future sought by the IRA are feasible. A workers' Republic of thirty-two counties may not be likely or around the corner, but it is a reasonable aspiration that has for six decades attracted the purists and for six decades been the proclaimed goal of the state. The contradictions between the ideal and reality have left a space for an armed conspiracy acting in the name of the nation.

It is clear that one of the major, often unstated, responsibilities of various Irish governments was the co-optation of nonconstitutional dissent—the legitimization of the state by real movement toward the ideal. When the process no longer worked, Dublin was in trouble and had to chose between the ideal and reality, adventure and honesty. The strength of the IRA eroded because most people will take half a loaf and will settle for the easy life. Secondly, the pattern of repression must be shaped to appear legitimate. For example, in Ireland emergency legislation, whatever the motives behind it, is a scandalous assault on all civil liberties. Yet with rare exception such legislation has been used solely against Republicans, and with relative restraint—when the IRA grows too arrogant. Hence much of the public understands and accepts the rituals or internment or imprisonment. Since Ireland is so conservative, there is almost no real radical Left; the only dissent is the Republican movement. No one else is penalized. After all, in a little country everyone knows just who belongs to the IRA and what they can legitimately do. Consequently, in Ireland, except for the radicals and libertarians (few on the ground), the emergency legislation is seen as legitimate, perhaps an exercise for narrow political purpose but not a danger to a free society. And it must be remembered that in Ireland a variety of institutions narrowly define just how free such a society may be. In sum, it would appear then that, when possible, accommodation pays and that, when necessary, the mix of repressive techniques must be regarded by the people as a legitimate exercise of authority.

So at the very end, over six decades, the Irish strategy of carrots and sticks—co-optation, coercion, faith, hope, and the aid of the bishops—has protected democratic government if not eliminated the idealists. There would have been less trouble if those in power had been more willing to talk about new realities rather than the old myths, but that is asking more than most politicians—Irish or otherwise—could willingly offer. For democratic governments elsewhere, the Irish experience with sixty years of armed subversion appears not to be readily transferable, except in demonstrating that carrots are effective, may cost less, and certainly taste better than the bitter dregs of police vengeance, unnecessary emergency legislation, and prison brutality. Apparently effective and swift accommodation and cunning co-option have a great part to play—certainly as great a part as coercion and repression, where less is most often more.

CHAPTER

11

THE ITALIAN EXPERIENCE: DEMOCRACY AND ARMED DISSENT, 1946–1977

Durante, et vosmet rebus servate secundis.

Aeneid,
Virgil

Unlike Ireland, Italy, until the nineteenth century a "geographical expression," had no deep democratic tradition. Before 1946 there had been only a brief interregnum between the monarchy and Mussolini, a time of chaos and uncertainty that gave birth to the Fascist regime. In the plebiscite on the monarchy in 1946, only fifty-four percent opted for a Republic—hardly a glowing endorsement for the new, largely unknown politicians who had lived their lives out of public sight in the underground, in exile, or in conventional professions. The times were harsh. The war had devastated the country. Cities were in ruins, communications disrupted, whole populations in want, the economy shattered. Some had faith, but few hoped for too much. As Virgil had written, "They endure for a while, and live for a happier day." On the other hand the Italians, unlike the Irish, did not anticipate that their sufferings, however prolonged, would lead to conquest.

To the amazement of all, not least the Italians and their leaders who wrought the miracle, Italy entered, if not immediately, the most happy days imaginable—not just for the rich, the well-born, and the well-connected, but also for large sections of the population as a whole. And from the first the Republic revealed a relatively static parliamentary pattern that assured a firm foundation for economic development, sparked in part by American aid and in part by native ingenuity.

The Christian Democrats alone or in alliance ruled from Rome. At the national level their greatest rival, the Italian Communist Party (PCI), could not hope for electoral triumph alone, and probably not in a Left alliance either. In fact the tendency over the next two decades was for the Left to be absorbed into the Christian Democratic center—first the Giuseppe Saragat Socialists (PSDI) and then the Italian Socialist Party (PSI) of Pietro Nenni shared power, lost the hunger of radical change by nibbling on the fruits of office, and left the Communists as the only significant opposition from the Left. On the Right and Center, a variety of small parties arose but could hope for little more than an accommodation with the dominant Christian Democrats.

It was on the far Right that there was one undeniable soft spot in the neo-Fascist *Movimento Sociale Italiano* (MSI). That the Mussolini era should attract the nostalgic, those who glorified in the old days, some ultranationalists, and the young eager to serve under stirring slogans was hardly surprising. Where the monarchists had faded away, maintained only briefly by charisma and habit, MSI managed to enlist new recruits and attract the old faithful—not enough to make a *real* difference but sufficient to maintain a real presence in the Chamber of Deputies. The MSI leadership insisted that the movement was not Fascist, which would have been against the law, but repeatedly slipped into Black Shirt rhetoric. And beyond the MSI there were strange fringe groups with romantic names who dreamed of a Fascist revival.

Some, like Prince Junio Valerio Borghese, the Black Prince, were silly anachronistic "aristocrats"; others were ultra-nationalist officers; a few were strange dreamers, known, if at all, only to the police. And a few, no matter what was said, appeared to be MSI members in good standing.

Elsewhere there would have been little concern over a small minority party and its lunatic fringe, but in Italy "Fascism" cast a long shadow. The Left was outraged that yesterday's criminals had resurfaced in the Chamber of Deputies. The Center usually chose to consider the MSI as a Right-National movement, not a Fascist one, and to discount the Marxist concern with the MSI as a political maneuver. The Communists, with twenty-five percent of the votes, vast resources, and friends abroad, could be a far greater challenge to a democratic system than a Fascist revival; but the PCI on the parliamentary stage, like the MSI, had no power. Politically, then, despite occasional Communist triumphs in local elections, Italy for two decades was safe and secure in the hands of the Christian Democrats and their allies. Italian elections became of concern only to specialists. There was movement from time to time, but not much. The PCI continued to do well locally, to control not just trade unions but also a whole array of workers' organizations, trade and economic associations, publishing companies, and insurance co-ops—a vast, Marxist vested interest in stability and growth. Not co-opted like the Socialists, the Communists—whatever their rhetoric and ultimate ambitions—gradually became very much part of the Italian political system. And the resulting political stability encouraged the real transformation in Italy: the economic miracle.

The figures alone of the Italian economy for the first two decades of the Republic were impressive. From a near-zero point, the growth averaged 5.6 percent a year. By 1968 the gross national product stood at 74.8 billion dollars—an increase of 370 percent since 1953. The figures, however, give little indication of the change in Italian life—a pedestrian could first afford a Vespa motor scooter, then a Fiat sedan, and finally drove a second car to a summer home. Two million Southern workers moved north to work in new or expanded industries. Billions of dollars were invested in the South and the islands. Vast, new high-rise districts spread out from the major cities as Italy became an urban country. Fiat and Olivetti became names to conjure with. Italy was no longer a land of pasta, *dolce far niente,* and long siestas. Italy was vital, growing, an integral part of the Atlantic alliance and the Common Market. Under the Christian Democrats, the country appeared stable and mature, the political system capable of accommodation, the economy growing, the more dreadful social inequalities passing, the future assured.

As the Republic moved into the third decade, however, the glow began to fade. The Christian Democrats grew older but not much wiser. The same names reappeared in the constantly changing cabinets. The central bureaucracy was huge and ineffectual. The education system was unresponsive. Unemployment was chronic. Despite fifteen billion dollars of expenditures, the regional development programs had not been a smashing success. The Southern emigrants in the North had not been adequately absorbed. The new wealth was not fairly distributed or even in some cases taxed. The housing shortage was seemingly permanent. There were repeated scandals and continuing incompetence—and ultimately no one was responsible.

But if there was concern, there was no real anxiety. Italy had never worked very well in the past; after twenty years, it seemed to work better than anyone could have anticipated. Growing pangs were to be expected, and Italy was not alone in running into developmental snags. Few could see devastating economic changes or a transformation of the electoral balance. Almost no one could foresee an extraparliamentary challenge to Italian democracy. In fact Rome had largely ended the only visible extraparliamentary "terrorist" challenge—the violent separatist movement in South Tyrol.

Separatism for geographical or ethnic reasons had been considered as a potential postwar Italian difficulty. During the war there had been maneuvers toward an independent Sicily; various areas, like Sardinia or Trieste, might resist rule from Rome. The most serious problem was the Alto Adige, South Tyrol, a maze of Alpine valleys snuggled up to the Austrian border and largely populated by German-speaking people. They were Austrians, cut off by the settlement of World War I. Fascist efforts to Italianize them failed and so had accommodation. In 1960 Austria had alleged unfair treatment of these former Austrian citizens on the wrong side of the Alps and raised the question in the United Nations. Soon thereafter the government in Rome opened negotiations with local leaders that focused on greater regional rights. Concurrently, militants seeking union with Austria opened a campaign of bombing, sabotage, assassination, and booby-trap attacks. Italian soldiers and frontier guards were killed and electric pylons were toppled. Rome suspected Austria was supplying sanctuary. Finally, in 1969 a complex package of concessions, the result of nearly a decade of discussion and two years of intense negotiations, satisfied, if barely, the vast majority in the region. There would be home rule and parliamentary approval from Rome and Vienna for the package. The militant irredentists had been deserted by Austria and now by the pragmatic in South Tyrol, many of whom in any case had doubts about the ultranationalist bombers. Rome had, if slowly, accommodated violent dissent. If not exactly solved, the Alto Adige problem went away.

During this period, if Rome noticed a "terrorist" problem, it was an imported rather than domestic matter—and there the Italian government appeared to some unduly lethargic about the responsibilities to international order as the country became an arena for terrorists and their enemies. In July 1968, after the PFLP hijacked the El Al 707 out of Rome and diverted the plane to Algiers, Italy became a magnet for terrorists, for their friends and enemies, the site of sabotage, assassination, and massacre—violence that had nothing to do with domestic grievance and which engendered seemingly only the most ineffectual technical and tactical response. The Italians felt, as did others, that the Palestinian confrontation was not *their* problem, not *their* responsibility. When the Israelis, as was their habit, pointed out Italian shortcomings in airport security, internal intelligence, and law enforcement, those responsible were outraged. Government indignation was hardly eased when others than the Israelis began to complain about Italian incompetence. Yet, in the five years after the PFLP hijack in 1968, there was only limited evidence that the Italians were taking serious remedial action. The Israeli Wrath of God carried out operations in Italy, the Palestinians seemed to be in residence, the transnationals sabotaged an oil refinery at Trieste, and airliners were repeatedly hijacked out of Rome.

The Leonardo da Vinci International Airport on the coast west of Rome had been built at great cost. Whatever its faults, and they were legion, any adjustment for security reasons could only be made at a considerable investment and real inconvenience, so nothing was done. Inspections were cursory and the "inspected" and others could still mingle freely. There was no inspection for transient passengers until they reboarded a different carrier, no serious security force, no effort to protect airplanes on the ground or keep them away from the terminal. Leonardo da Vinci Airport was a terrorist dream.

In September 1973, Palestinian fedayeen tried to attack an El Al plane with a rocket launcher but failed. Three months later the continuing vulnerability of the airport became horribly apparent. On December 17, 1973, five young men arrived on a flight from Madrid. They made their way to the transit lounge. They were quite unremarkable—tall, dark, each with a mustache and hand luggage—except they were a Palestinian fedayeen operational group. As soon as the five reached the security barrier, they stopped, pulled submachine guns from their hand luggage, and fired on the crowd clotted around the gate. The plate glass walls of the terminal shattered, spraying shards over the stunned passengers, who began to scream and scramble out of the way. It was Lod revisited. The fedayeen swiftly rounded up Italian security men, killing one with a burst of machine-gun fire when he attempted to escape. Two of the terrorists ran forward and tossed incendiary grenades into a Pan Am 707. The passengers were trapped in the swirling smoke and spreading fire. Meanwhile, the other three fedayeen took over a Lufthansa 737 and waited for their colleagues. Once all five were aboard—leaving behind them thirty-one dead on the ground, mostly aboard the flaming Pan Am jet—the pilot was ordered to take off. They then embarked on a typical terrorist tour—stopping off at Athens, where they killed one hostage, and finally landing without permission in Kuwait.

There was as always outrage at the terrorist drama from all quarters, and the Italian government came in for some criticism. In response the government did tighten up security: guards armed with automatic weapons appeared inside the terminal and on the tarmac, and vulnerable jets were parked way out on the tarmac, far from the boarding gates, and served by buses. Leonardo da Vinci Airport was no longer a terrorist dream and the Italian government no longer lethargic about the problems of international terrorism; in fact, over the course of a violent decade the Italian government seized on the threat of the new terrorists—but Italian terrorists, not transnationals.

A Decade of Violence, 1967–1977

The first sign of trouble came in 1967, a year before the Paris May Days, when the Italian secondary school students rioted in protest at the unresponsive, archaic educational system.

Unlike the riots of gesture in Paris, the Italian riots were smashed swiftly and without dialogue. Opposition to the system went underground and the student opposition in 1967 and 1968 began to use the language of the Left. As time and other priorities whittled away the activists, those who demanded radical political and societal change, only the truly dedicated remained. They were, however, without a specific direction or an appropriate agenda, with only a posture on the far Left of the political spectrum, beyond the Communists, beyond parliamentary politics, and, like the lunatic fringe of the Fascists, beyond notice. In the late 1960s the Italian advocates of violence, ideological ultras, were quite unknown. Then, between April 1968 and April 1969, seventeen "symbolic" bombs exploded in Italy. In August 1969, bombs detonated simultaneously on seven trains in northern and central Italy, injuring twelve people. No one seemed to know who was responsible—perhaps "anarchists"—and no one seemed greatly to care. Then, on Friday, December 12, in Milan, a bomb exploded in front of the Banca Nazionale dell'Agricoltura, killing sixteen people and wounding ninety. Bombs had also gone off simultaneously in Rome. The total of 16 dead and 140 wounded could not be ignored. The country was shocked. Within days, 244 persons were questioned and 55 held, but, from the very first, anomalies dogged the investigation. The police first charged Pietro Valpreda, a known anarchist, with complicity. Valpreda insisted he was being framed. Eleven more anarchists or members of extreme Left-wing political groups were "believed to be involved." Then the police announced that one of the suspects, Giuseppe Pinelli, had jumped to his death from the fourth-floor office of Dr. Luigi Calabresi, head of Milan's police political bureau. Pinelli's suicide, according to the police, "bore all the signs of self-accusation." Although an anarchist "plot" had been a satisfactory explanation to nearly all varieties of Italian political opinion, the rush to close the case after Pinelli's death seemed strange. As the formalities dragged on month after month, other oddities occurred. Witnesses began to die. Armando Calzolari, a member of the Right-wing group associated with Prince Borghese, disappeared on Christmas morning 1969. He had been making accusations of murder against his colleagues. His body was found early in January. Another Right-wing lawyer, Vittorio Ambrosini, died—a suicide. He had voiced suspicions about the Milan bomb. Three anarchists, two of whom were character witnesses for Valpreda, died in an automobile crash. There were doubts about the prosecution's chief witness, a taxi driver who claimed to have taken Valpreda to the bank. In July 1971, the driver died, apparently naturally. The Anarchist Plot began to come apart.

Apparently, political violence had come to stay. In July 1970, the Rome-Sicily rail tracks were sabotaged. Six people were killed and one hundred injured when the train crashed. Symbolic bombs continued to detonate. There was arson at the Pirelli rubber plant in Milan. Clandestine broadcasts called for revolution. There were rumors of Fascist plots. Only very slowly did a picture of those responsible emerge. There were vested interests who preferred there be no "terrorists." In 1969–1970 the emergence on the Left of the *Brigate Rosse* alarmed the orthodox Communists and Socialists. Advocating extraparliamentary action and heaping scorn on conventional politics, the young people of the *Brigate Rosse* had lost faith in the system. The Right was delighted at such evidence of Left-wing violence. The Communist *L'Unita* was less delighted, claiming the *Brigate Rosse* people were pseudorevolutionaries or, more likely, provocateurs. In March 1971, after a murder and robbery in Genoa, the police discovered another ultra-Leftist terrorist organization of "Maoists." In Milan police found evidence of the *Brigate Rosse*'s part in the Pirelli arson case.

Then on March 17, Parliament was told that an arrest warrant had been issued for Prince Borghese, charging him with political conspiracy and attempted armed insurrection. On December 7, 1970, Borghese's neo-Fascists had begun, and then called off, a coup, code name Tora Tora. The Black Prince fled to Spain. Four others were arrested. The Communists were delighted. The evidence began to pile up that the neo-Fascists were deeply implicated in much of the violence, most particularly the Milan explosion in December 1969. The police began arresting Fascists who had left the "too moderate" MSI in 1956 and formed their own New Order. Thirty-four were charged with attempting to reconstitute the Italian Fascist party. There were rumors of other and more unsavory groups—including a weird Nazi-Maoist cell. The Black Prince's silly plan to march on Rome might be discounted; but many in the New Order and the wilder fringes of the new Fascism claimed friends in the army and in high places. And retired Lieutenant-General Giovanni De Lorenzo, accused by the Left of organizing his own coup in 1964, now marched with MSI.

Despite the odd bomb and the regular demonstrations organized by the Mussolini Action Squads or the New Order, the most serious challenge was perceived by the Center and the media to come from the Left. In early March 1972, the *Brigate Rosse* kidnapped an executive of a Milan electronics firm but released him several hours later. Then the wealthy Left-wing publisher Giangiacomo Feltrinelli was killed in a dynamite explosion. The Left claimed he had been murdered by the Fascists. The circumstances of his death remained obscure, but the investigation revealed a Milan center of "Partisan Action Groups" with detailed plans for urban guerrilla war. The groups had contacts with *Brigate Rosse*. The Communist Party insisted that the "radical net" had been stage-managed, timed to influence the general elections. Then, further arrests of neo-Fascists indicated that they had been responsible for "anarchist" operations and had, apparently as provocateurs, infiltrated far-Left organizations. Two were suspected of the 1969 Milan bombing. Spokesmen for MSI immediately disclaimed any involvement. Only the Christian Democrats, stressing the threat from "opposed extremists," seemed likely to benefit from the revelations concerning the Right or Left.

The next elections, however, revealed little change in CD strength (267 seats in the Chamber, instead of 266), but a significant gain for MSI (from 24 seats to 51), and a slight general move to the Right with a more conservative cabinet. There was no sign of an easing of political violence. On May 18, Dr. Luigi Calabresi, blamed by the Left for the death of Pinelli after the 1969 Milan bombing, was murdered outside his house. A vast security sweep throughout the country produced 230 arrests but no culprits. The street fights, attacks, and bombings continued. Giorgio Almirante, leader of MSI, indicated—perhaps—in a speech in Florence that a takeover might be necessary. The streets filled up again with demonstrators and there were more attacks on MSI officers and subsequently on Left-wing offices.

In a sense the continuing violence appeared the least of Italy's problems. Despite the protests, the government was as unresponsive as always. There had been no change in the educational system, only further repression. Essential services seldom worked properly. Nothing was on time. Any request generated endless paper work and no one had ultimate responsibility. There were always strikes but no subsequent visible gains for the workers, only inconvenience and a stagnant economy. There was a small, constant erosion of faith in the system's ability to adjust. The Christian Democrats had dominated the national government for a generation. They had grown old and arrogant but no more competent. The Opening-to-the-Left had brought a share of power to the Italian Socialist Party and with it the corruption of militancy. Some local and regional governments were effective, especially several Communist-run showplaces, although at considerable cost; but Italy was badly governed.

The year 1972 ended with riots protesting the third anniversary of Pietro Valpreda's arrest for the Milan bombing. Parliament had passed a law granting him constitutional liberty, a form of bail. The new year began with bombs; on January 17 alone, over two hundred petrol bombs were thrown at symbolic targets. In Milan, on May 17, Gianfranco Bertoli, an "individual anarchist," tossed a grenade into the crowd waiting for the unveiling of a bronze bust of the murdered Luigi Calabresi. A young woman was killed and forty-five people injured, five critically. The bearded, middle-aged Bertoli was a notorious character with a large "A" for Anarchist tattooed on his right forearm. No one believed that he acted alone. Interior Minister Mariano Rumor, who had just left the ceremony, suggested that Bertoli might be part of an international plot. The Communist *L'Unita* blamed the major American labor organizations.

On May 24, the Chamber of Deputies lifted the parliamentary immunity of the MSI leader Almirante so he could be tried for reorganizing the Fascist Party. Almirante insisted that he was innocent and that his party completely disavowed the fringe groups. His protestation apparently had some basis, for in November four people, the Compass Rose Group, were arrested in La Spezia for organizing a Right-wing coup. Part of the evidence was an assassination list of 1,617 names, including such obvious targets as Communist Party leader Enrico Berlinguer but also Almirante. The police suspected Prince Borghese might be behind the plot but in the meantime began arresting known neo-Fascists. There were more arrests in January 1974 related to the Compass Rose plot. Retired General Francesco Nadella and a major were jailed. There were rumors of other officers involved. In March an investigating judge ordered three neo-Fascists to be tried along with the anarchists for the 1969 Milan bombing. Italy now faced the ludicrous prospect of the most violent of political opponents, anarchists and neo-Fascists, being tried for the same offense. There were increasing revelations of Fascist violence, of strange organizations like the New Order, the National Vanguard, Mussolini Action Squads, the Steel Helmet, and of new charges and new plots.

As if to counterbalance the growing focus on the Right, on April 18, in Genoa, five armed men kidnapped Mario Sossi, the Deputy Public Prosecutor. *Brigate Rosse* demanded the release of several political prisoners. Sossi, "a fanatic persecutor of the working class," would be held in a people's prison and tried by a people's court. Although kidnapping had increasingly become an Italian cottage industry, the implication that Sossi might be "executed" unless eighty *Brigate Rosse* prisoners were released was a new technique. Sossi's superior, Francesco Coco, was unwilling to meet the terms and delayed proceedings. The days dragged on with no resolution, no releases, and no Sossi. On May 23, the *Brigate Rosse* finally freed Sossi. Coco announced that Sossi had been mistreated and so the prisoners were not released. The Sossi case disappeared from the front pages in less than a week—the violent pendulum had swung to the Right again.

On May 28, an explosive device stuffed into a garbage sack near the rear of the crowd attending an anti-Fascist rally in Brescia detonated. Six people were killed and another ninety-four injured, one fatally and many seriously. Everyone was horrified, but no one really believed that this was simply an isolated "incident." All spring there had been a series of bombs claimed by the new Black Order. In 1973 the New Order, founded in 1954, had produced a journal *Year Zero,* initials often used in proclamations of responsibility. In April 1974, New Order-Year Zero had been transmuted into the political movement of the New Order and then into the Black Order. According to reports, there were perhaps only a few hundred members in the movement. If the Black Order really was so small, there were many who felt that it had a great many associates or sympathizers and considerable resources. The Fascists appeared to be orchestrating a *strategia della tensione* that would create conditions appropriate for the establishment of an authoritarian government. If so far the Black Order had not exactly acted with impunity, then still the Italian security forces left something to be desired.

Italy had a paramilitary force of 80,000, a public security force of 80,000, and a finance guard of 40,000, along with a military intelligence force of 2,000, for a total of over 200,000—one security officer for every 275 Italians. And the key, supposedly, was the elite *Servizio Informazioni Difesa* (SID); the military counterintelligence force, (SIFAR) had been reorganized under criticism, not the least of which was the Left's claim that director General De Lorenzo, had planned a coup in 1964. The new SID had run into trouble in 1969 for some questionable practices related to dossiers. There had been further reorganizations, but rumors persisted about SID. There had been a strange security alert in Rome in January 1974 and no subsequent explantion. Fascist complicity in the Milan 1969 bombing was now accepted, and by 1974 there were also strings leading toward SID. There were questions beginning to be asked about Director General Vito Miceli and indications that he might be transferred. The government announced that the police chief of Turin, Emilio Santillo, had been appointed to direct the new *Inspettorato anti-terrorismo* (Anti-terrorist Office).

On May 30, police killed one member of a Mussolini Action Squad in a shoot-out. More arrests produced more documents. There had been plans to machine-gun a labor demonstration, to seize a Carabinieri barracks, to blow up bridges, and to assassinate prominent politicians. On June 17, two MSI officials were shot and killed in Padua. Almirante blamed the Left. On June 24, his colleague Admiral Bino Birindelli, President of MSI and an old-line nationalist, resigned, uneasy over the leather-jacketed street fighters and the rumors of plots. Early in July, Defense Minister Giulio Andreotti removed General Miceli, now implicated in the Borghese coup and the Compass Rose conspiracy, and appointed Admiral Mario Casardi as head of SID. Rumors about Miceli grew more precise and evidence of Fascist plots more accurate. It was announced that President Giovanni Leone had been the intended victim of an assassination plot. Then came another Fascist outrage. On August 4, a bomb detonated on the *Italicus* Rome-Munich express just outside Bologna. Twelve people were killed and forty-eight injured. A typed message from the Black Order, left in a Bologna telephone, took responsibility for the blast. There was the usual round of vengeance bombs, cordons and searches, arrests, and revelations. SID was under increasing scrutiny. There had been some sort of warning given by Almirante about the train bomb that had been muddled in SID. There had been the mobilization in Rome in January, now made public but still unexplained; and there had apparently been another go-round in August, but no details were released.

The most spectacular development of all was the arrest of General Miceli on October 31 in Rome. He was accused of complicity in both the December 1970 Tora Tora coup of Prince Borghese, who had died in Spain in August 1974, and the 1974 coup plot. On December 14, General Ugo Ricci was arrested. The former Chief of Staff of the Air Force, General Duilio Fanali, was under suspicion on similar charges. Further Fascist armed nets were uncovered. There were rumors of a secret international meeting of Fascist parties in Lyon in December. In Italy the year ended with further street fighting between Left and Right extremists—in Rome three policemen were shot and another badly burned on December 22. Everyone's worst fears seemed justified. There were real terrorists, Left and Right; real plots; real, if unsuccessful, coups; real treason within the security forces, as well as demonstrable incompetence. Worst of all, there were street violence and bombs, symbolic and deadly—all the techniques of urban guerrilla war had become part of Italian life.

In 1975 another organization surfaced on the far Left. In Naples in March a premature bomb had brought in the police. They discovered the *Nuclei Armati Proletari* (NAP), an ultra-Left group of uncertain origin. On May 6, Judge Giuseppe di Gennario was kidnapped for what the police at first announced were not political reasons. This proved wishful thinking. NAP announced three days later, during a prison revolt in Viterbo, that they held Judge di Gennario, an expert on prison reform. He had actually been kidnapped because he was progressive rather than reactionary, and the focus of NAP's demands was the prison system. Not unlike certain American groups, the NAP people had become involved in "politics" in prison or in association with prisoners and saw the criminal justice system as one more unresponsive, brutal institution. NAP's rather limited demand for legal aid for their members was met in a broadcast by the state radio and television. Di Gennario was subsequently released.

In June the police shot and killed Margherita Cagol Curcio, wife of a leader of the *Brigate Rosse*. She had been found in a farmhouse that was a secret prison for the wealthy vermouth manufacturer Vittorio Vallarino Gancia, kidnapped the day before. The police also announced that seven NAP apartment hideouts had been uncovered and six people charged. Margherita Curcio's death posed a difficult problem for the orthodox Left. Their explanations of the *Brigate Rosse*-NAP violence had been that it was a conspiracy organized by the Right—and probably the CIA—manipulating innocent idealists to heighten tensions just before the elections. Whether it was true or not, it was convenient. The Curcios, however, had impeccable Marxist backgrounds, had belonged to the appropriate organizations before their shift to revolution, and could be neither dupes nor innocents. *L'Unita* grudgingly admitted that some of the *Brigate Rosse* might not be provocateurs on orders from Italian or foreign conspiratorial centers and might even be acting in good faith.

The events of 1975 had destroyed the various comforting myths about the nature of Italian political violence. In January 1975, the trial arising from the Milan 1969 bomb incident was postponed five years. It was clear there had been no "anarchist" plot, that it had been a Fascist bomb, and that there had been official complicity. The evidence had been so juggled and maneuvered that a plain tale of events seemed impossible. In March the Milan prosecutor announced that Feltrinelli had been involved in Leftist urban guerrilla plots. His act of indictment charged forty-seven members with a conspiracy to overthrow the country's institutions by the use of armed force. Coupled with Margherita Curcio's death, the orthodox Left could not claim violence was all a reactionary plot. Nor could the Right blame "the anarchist Left". On November 5, seventy-eight people including General Miceli were ordered to stand trial on charges connected with Borghese's Tora Tora coup. The Black Prince might have been foolish, but any plot involving the director of military intelligence was not simply a puerile exercise.

The next, and for many most distressing, plot was revealed in January 1976. In 1972 American Ambassador Graham A. Martin, over the violent opposition of the CIA, had given the same General Miceli payments of 800,000 dollars for a propaganda exercise that would call for no subsequent accounting. The CIA had insisted that Miceli was associated with antidemocratic elements and that the money would be wasted. Payments by various foreign patrons to Italian political organizations were common, although the Soviet methods were more subtle. For much of the orthodox Left, the CIA was the major culprit, all but manipulating the "reactionary" forces, so that the revelation that the CIA had opposed Martin's initiative came as something of a surprise. Equally unpleasantly surprised were those in Italy and the United States who discovered that the accusations from the Left had substance—Washington *had* been underwriting "undemocratic elements."

More evidence concerning undemocratic elements came in March, when the former head of the defense section of SID, General Gian Adelio Maletti, was arrested and accused of funding Fascist infiltrators and provocateurs and of creating parallel structures inside SID responsible only to him. Italian security was a complete disaster area. Serious attempts were underway to correct the SID abuses; Santillo's Anti-terrorist Office had been ferreting out suspects. Twenty members of the *Brigate Rosse* were on trial in Turin, and arrests of neo-Fascists continued. What also continued was the escalating violence. Still another political crisis loomed.

Regional elections had revealed a substantial shift to the Communists and consequent losses by the Christian Democrats. There was real evidence that new elections might force a historical compromise, a government of the Communists and Christian Democrats. Some even felt the shift might produce a Left government, excluding for the first time the Christian Democrats. No one was quite sure to whose advantage an early election might be. It appeared that the Italian Socialist Party, sensing a Left tide, was eager for a new test; and so, as the prospects of elections increased, the militants took to the streets once more. Italy entered a violent spring.

In March and April there was a growing decay of order. There were Molotov cocktail attacks, police chases and shootouts, gunfights in the streets, widespread arson, firebombs, and assassinations. There were also the small incidents—arson in classrooms, random sabotage, cars wrecked in the streets, robberies—all in the name of the struggle against the system. In April demonstrators rushed through the streets of Milan smashing expensive cars, symbols of "the system." Violence had become endemic.

Finally, Aldo Moro's minority Christian Democratic government ran its course. There would be elections in June. On May 28, south of Rome at Sezze Romano, MSI held a rally that collapsed into violence. Sandro Saccucci, a member of the Chamber with a long record of provocative violence, was sought for murder after a young Communist was killed in the melee. Saccucci disappeared, then surfaced in London, where he was arrested by Scotland Yard and held for extradition proceedings. Almirante withdrew Saccucci's MSI membership. Parliament withdrew his immunity; but many felt it was too late. MSI had been disgraced and the Christian Democrats were the gainers. On June 8, in Genoa, Public Prosecutor Francesco Coco left his car and driver and began walking towards his home for lunch. Several young men suddenly appeared. The driver was shot and killed. The bodyguard was shot and killed. And Coco was shot and killed. The following day in court in Turin, Prospero Gallinari, in the name of *Brigate Rosse*, accepted responsibility for the death of Coco. If Saccucci had cost the Right a million votes, then the assassination of Coco seemed likely to cost the Left another million.

No one, of course, could accurately judge the impact of the two murders on the electorate; but on the morning after the final polling, it was clear that the country had been polarized. The Christian Democrats kept their place as the largest party, with approximately the same number of seats. The Communists made huge gains. The smaller parties were decimated in most cases, and the ambitious Socialists were sorely disappointed. A monocolor, single-party, minority Christian Democratic government emerged—made possible by Communist abstention and ultimately cooperation, a most radical innovation. The long period of postelectoral maneuvers brought some considerable easing in the random violence. Then, on July 10, Roman Public Prosecutor Vittorio Occorsio was machine-gunned while sitting in his car. Fascist New Order pamphlets were scattered over his shattered body: he had been "guilty" of serving the democratic dictatorship and persecuting New Order militants.

And there was no end in sight. In the spring of 1977, university students, embittered at certain unemployment on graduation and the loss of their student stipend, angry at the huge enrollments, unresponsive regulations, small and often part-time faculties, and archaic facilities, took to the streets. On March 5, 1977, in Rome, Fabrizio Pansieri was sentenced to nine and a half years for the murder of a Greek neo-

Fascist student, two years before. The radicals of the Left mounted protest demonstrations. On the following weekend, Friday evening, March 11, in Bologna, one thousand students turned out to protest the sentencing. They represented a variety of radical organizations, including the small Left-wing parties beyond the Communists, like *Lotta Continua* (Continuing Struggle) and *Democrazia Proletaria*—and, of course, those just out for excitement. The demonstration collapsed into a riot. The police opened fire and killed Pier Francesco Lorusso, a medical student at the University of Bologna and a leader of *Lotta Continua*. Radical students throughout Italy were outraged and the result was a weekend of rampage and violence. Students in major cities—Turin, Milan, Florence, Naples, Rome—ran amok, tossing firebombs, ransacking stores, and attacking police. In Bologna police moved in armored vehicles. The city center, with swirling clouds of gas, squads of police carrying rifles, and masked demonstrators, often armed, became a war zone.

The most violent display came in Rome on Saturday. Students flocked in from all over Italy—many bringing Molotov cocktails and riot equipment. On Saturday there were fifty-thousand demonstrators—most out to disrupt the city. The headquarters of the Christian Democratic Party was firebombed. Gangs of radicals wearing masks and carrying iron pipes raged through the center of the city; forcing drivers from automobiles, they created burning barricades of Fiats. Storefront windows were smashed; no traffic moved; police and rioters exchanged fire.

On Sunday morning the center of the city looked like a battleground—gutted cars, ruined stores, the streets littered, and everywhere the lingering odor of burnt rubber and gas. The government banned further demonstrations. Apparently, the riot had run its course—a dozen police wounded, one, Giuseppe Ciotta, fatally shot in Turin by an unknown "Fighting Brigade," and the authorities everywhere stunned. In Rome Interior Minister Francesco Cossiga announced that the government would not tolerate further violence, that there had been a plot to spark urban guerrilla warfare. Massimo d'Alema, the secretary of the Communist youth organization, condemned the violence and announced that Italy faced a new phase of tension and provocation that might throw the country into a dramatic crisis.

The PCI to its horror found the students paid no attention to their advice—the Communists had become part of the system they hated. None of the conventional parties failed to condemn the rampage: violence from the Left or Right would be unacceptable. The students, the radicals, the real or potential urban guerrillas, drifted out of sight—obviously unconvinced by the protests of those who had gone soft within a corrupt parliamentary system. The walls of Bologna, long a Communist-run showplace, were covered with the graffiti of the March rebels: "We want everything. We want to destroy everything." The government was shocked, many observers were as well—fifty thousand radicals had turned the center of Rome into a war-zone on twenty-four-hour notice, in the name of nothing at all, simply "to destroy everything." Italy—increasingly troubled economically, with a government dependent on Communist toleration, with accumulating ills and no sign of a will to act—had been revealed in the turmoil of the streets on the edge of crisis.

That at last the scope of the crisis was recognized did not indicate an early solution. The disorders continued, and a new technique of violence appeared on the Left: conservative journalists, Christian Democratic politicians, and officials in major industries were shot in the legs—*azzoppamento* (laming). By mid-July 1977, thirty-three men had been wounded by the *Brigate Rosse* in these *gesto esemplare* (exemplary deeds). Yet inside and outside prison the total numbers involved remained small. In July 1977, there were 263 Leftists under detention in investigations of 37 murders, 13 attempted murders, 48 robberies, 26 kidnappings and other crimes, while 343 Fascists were in jail in connection with terrorism and violence. Still, the assassination of judges, the sabotage of Northern industries, the bombs and shooting, the trials of high security officials involved in coups, and the tens of thousands of rampaging university students—all indicated that the "few" were symptomatic of a general societal *angst*. Italy might no longer have an unresolved nationality problem, but Italy might well no longer be an efficient democracy—after all, it had happened once before and there had been a march on Rome and an end to liberty in the name of authoritarian order.

In two areas the Italian response to political violence—international terrorism and ethnic separatism—has been almost conventional. In the case of the international terrorists largely out of the Middle East, Rome only slowly accepted the bombers and gunmen as an Italian problem, and gradually and not very effectively deployed security techniques and tactics, and then participated in regional and international covenants. Each incident tended to be handled on an *ad hoc* basis, with the new antiterrorist squad no more in control than similar units elsewhere with a mixed record of concession and repression. The response to the separatism of South Tyrol was an undeniable success, the devolution of real power from the center, a process that continues in other regions (most recently with further extensive grants of local control through parliamentary legislation—supported enthusiastically by the PCI in a legislative pact). It is, however, the challenge from the entire spectrum of extreme ideological dissent that makes Italy almost unique. In Ireland there has been no international problem, no dissent from the far Right or the far Left, only from the traditional Republican movement seeking more for the nation than the nation seems to want. But in Italy there is something for everyone: Fascist squads; the ultra-Left *Brigate Rosse;* the intellectual-prisoners alliance of the NAP; coups plotted by generals and aristocrats; coups plotted by the middle class and majors; foreign interference; betrayal in high places; and even a Nazi-Maoist splinter group. And the Italian response to this wealth of violent dissent has been equally various.

Basically, none of the opponents of the state has a monopoly on the kinds of assets possessed by the IRA. The Fascists might claim a descent from the true nation corrupted and sold out by the government, but no one outside MSI would buy the proposition. The ideals of Fascism or the far Left may have takers, but these organizations represent deviant currents, not the mainstream of Italian political history. Their rituals and ideals are for the few, not recognized by the many. None has an accepted mission. No one outside the faithful sees them as contenders for power, capable even of a veto such as the IRA has in Northern Ireland or even of "bombing down" existing institutions as the IRA claimed in the case of the Northern Ireland government. In fact, it is not so much that the militants of NAP or the New Order have assets to wield, along with their perceived truth and small arms, but rather that the state has so many unpaid debts—debts falling due.

For much of the violent decade, only gradually did the inadequacies of the system become apparent, especially since those excluded by the system were inarticulate, without leverage or visibility. Students, on the other hand, are nothing if not articulate and, in the midst of riot, highly visible. The responsible in and out of parliament, however, did not perceive the growing domestic violence as significant or relevant to serious political matters. In order for the events to make sense, it must be understood that almost without exception no one within the Italian system or with responsibility wanted political violence to be rational acts of rational men. When there was "violence" it was discounted, ignored or blamed on traditional enemies—the Communists blamed the Fascists or the CIA or the Christian Democrats; the Christian Democrats blamed the Communists or unknown conspirators. There might be a conspiracy or foreign plots, but there was not a *serious* political problem. Not, in fact, until the spring of 1977 was the relation to the ultra's violence and real, substantive issues accepted.

One of the difficulties for the conventional in accepting the dangers of political violence, other than the partisan benefits it might bestow, was simply that the gunmen seemed both alien and irrelevant. The Communists might like to worry in public about the neo-Fascists, but in private no one took the posturing and provocation very seriously. The conservatives might find a revolutionary Red Plot comfortable fodder, but no one believed the Maoists were really at the gate. And the advocates of violence seemed so unrealistic. Both the ultra-Left and ultra-Right apparently really believed that violence would bring down the system. The Left activists, like their colleagues of the Baader-Meinhof Group in Germany or the Angry Brigade in England, felt that a campaign of violence would rally a bemused and manipulated working class behind the revolutionary banner. In the case of the ultra-Right, the model was either a coup-conspiracy or a strategy of tension that would undermine parliamentary government and ease the way for men of authority to save the state. The ultras shared a fervent opposition to electoral politics—a sham, a facade to cover an exploitative and corrupt system (the ultra-Left) or a dishonorable and shameless betrayal of the national ideals (the ultra-Right). Both insisted on radical changes that opposed the ideals and aspirations of all the conventional parties. The extraparliamentary militants at both ends of the political spectrum sought to kill the old system. By any measure, the neo-Fascists were potentially more threatening. Their "cause" rested on recent historical foundations, attracted a variety of talents, skills, and resources. It could perhaps, involve still more recruits if there were the appropriate momentum, and could to some degree count on the overt MSI. The covert Left had but a few friends (almost none in parliament), the limited resources of the young and dedicated, and a luminous ideological vision. In both cases, however, the ultras soon resorted to violence, extortion, theft, riot, arson—intimidation to further their own cause and maim that of their declared enemies. For both, violence and counterviolence were seen to pay immediate dividends. The Fascists were more optimistic, more ambitious, more certain—they, after all, had more assets and more friends in power; but both assumed that the center, the system, was doomed. But the system did not feel doomed; in fact, it only gradually came to feel threatened. The real threat, however, was not in the power that could come out of a gun barrel but in the powerlessness of the center seemingly unable to effect events or determine the future. If an effective democratic state without a submerged nationality problem does not have to worry about a terrorist threat, this is not the case with an *ineffective* democracy. There must be a capacity to accommodate legitimate demand for change, even when the request is riotous. The students rampaging through the streets of Rome do not want a Marxist millennium or a Fascist Eden—most do not even know what political future they might prefer. What they want is *more*—a piece of the action, a decent university, a promising career, a taste of the good life. Some may not want to work for this, but most would. And the government has not been able to give them more, not kept promises made to them, not been able to repress them at this most volatile time of their lives—has in fact been grossly inefficient. Worse, with a fragile government any reform is certain to cause increased turmoil. The University of Rome cannot function as a university with a student body of 135,000 and a small, threatened faculty. There must be fewer students, the discarded certain to protest in the streets, fewer acceptances, the denied will protest in the streets, and a larger more effective faculty, just at a moment of insolvency. And even then if the nettle is grasped, there must be jobs after graduation. In sum, the state must efficiently cope with legitimate dissent or, in delay, see its assets of legitimacy decay and its capacity to act, even repressively, diminish. Its very democratic system may come under suspicion as the appeals of *any* alternative to institutionalized incompetence beckon. And in Italy it all happened once before, as the Fascists know quite well—as *every* Italian knows.

So far the Italian gunmen represent no one but themselves or a tattered tradition. So far the native terrorist threat is hardly lethal—after all, Italy has never worked very well. So far those who have taken to the streets or sabotaged their factories or thrown stones at the police are not committed to revolution. This may not be apparent to those who equate democracy with turmoil, who insist on absolute order, who doubt the will of the government—or the state as so constituted—in the face of open rebellion. The odds are that Italy with scrape through, but it will not be with a mix of techniques and tactics, new antiterrorist squads or electronic devices, more repressive legislation or more police in the streets. Italy faces not so much a terrorist problem as a need to restructure its society—a strategic response—so that the conventional assets of legitimacy can be maintained, so that the center will be immune to the challenge of the frantic, the few who can no longer be co-opted, who want not more but everything—and often everything destroyed.

PART FOUR

Conclusion

Dopo il fatto, il consiglio non vale.

CHAPTER 12

AN APPROPRIATE RESPONSE

After a decade of dismal terror, there can be few left who are still innocent of the new politics of atrocity and the war waged by tiny "armies" of fanatics bearing strange devices. All now know the long and grotesque litany of massacre: Lod-Munich-Khartoum-Rome-Athens-Vienna. Now millions are familiar with the luminous dreams of the obscure South Moluccans and the strange Japanese Red Army, with the fantasies of the Hanafis and the Symbionese Liberation Army, and with the alphabet of death—PFLP, FLQ, IRA. Carlos-the-Jackal is a media antihero, and Croatia is now found in the headlines instead of in stamp albums. Anyone can be a victim, can ride the wrong airline, take the wrong commuter train or accept the wrong executive position abroad. While opening mail, passing a foreign embassy, standing in an airport boarding line or next to a car, or attending a diplomatic reception, any of us may draw a "winning" lottery ticket in the terrorist game. And everyone is the target of the television terrorists, choreographing massacres for prime time. After each crafted incident, terror still produces intense anguish and indignation, a plea if not for vengeance then at least for effective action. The target-audience has not become inured to violence. Repetition has established ritual, not ennui. Sophocles never pales nor, so far, has the murder of innocents, brought to us personal and close-up by the media.

Terror thus has become almost institutionalized, an integral aspect of a Western society unable to accommodate all political aspirations, a society whose very stability and success have driven the deprived and desperate into violent gestures that leave innocent, blood-soaked victims strewn across public floors. For the terrorists there are no innocents. What matter the victims, provided the gesture is beautiful? And for the West these macabre gestures have provoked an uncertain and shifting response that reflects the predilections and contradictions of the specialists and the responsible. If only terror could be presented to the public as a natural calamity, deadly but irrelevant to the major current of national life, then the general anguish would erode and the plea for "action at any cost" would evaporate. Hurricanes actually *do* kill far, far more than the terrorists. But it is not the number of broken bodies or ruined homes that matters; it is the perception of horror wrought by men who are beyond reason. How can one reason with General Field Marshal Cinque or Carlos-the-Jackal? How can one reach an accommodation with those who kill in the cause of earrings or seek a union with their victims in the stars? Unfortunately, revolutionary terrorists are not viewed by the public as an unavoidable calamity but as a challenge that must be met and overcome at great cost—even, some feel, by severely narrowing open societies. The public and many in positions of power simply will not buy the argument that terrorism kills relatively few. The many feel political terror cannot be ignored or tolerated and must be met with the resources of the challenged societies, for to do less might open the door to anarchy.

Few are interested in mere body counts. The fact that more people are regularly killed in auto accidents on the roads of Northern Ireland than in terrorist incidents does little to ease the anguish of those who walk the bombed streets of Belfast or past the burned out country pubs. No one anticipates death on the roads, but everyone in Ulster waits for the crunch of a car-bomb or the crack of a sniper's Armalite. The numbers, charts, and graphs cannot erase the fear of a danger that is wanton, awesome, random, and alien. The fatalities produced by the automobile can be tolerated; even arguments that the reduction of speed limits *guarantees* a reduction in deaths and injuries persuade very few to accept a limitation on their freedom to drive. Here convenience, not risk, is the priority. But in the case of terror, each death becomes a direct affront to normal life. The revolutionaries have no *right* to murder the innocent in the name of strange gods, to blame the system, not themselves, to kill and go free. So, although a most effective response would be to ignore the terrorists, raise some hurdles, concede when necessary, accept such violence as a natural and not very important aspect of life, unpleasant and unsavory and unavoidable, this is not possible. The public are not "terrified" at each massacre. They are outraged and indignant. After ten years this is the first and basic constant that any response to terrorism must accept.

Obviously, not all revolutionary violence is socially irrelevant: some terrorists are, indeed, signs of bad times, symbols of real sound and fury. Often the gunmen reflect severe and all but insoluble societal dysfunctions. In Ulster, still a zero-sum game in which every gain for the minority is seen by the majority as its loss, the gunmen represent something other than their own whims and fantasies. After World War I the Frei Korps in Germany and the Black Shirts in Italy were not simply self-proclaimed armies of redemption. In Spain in 1936, the disturbances and killings in the streets represented real threats. Yet the principle remains: efficient, democratic countries without an unresolved nationality problem need not fear domestic revolution from terrorism. Riotous behavior by those demanding radical change, symbolic bombs, psychotics, and at times, briefly, the unrepentant gunmen like those of the Baader-Meinhof Group or the *Brigate Rosse* can appear, but these are a *normal* aspect of an open society which guarantees disorder and chaos.

In some societies like those of Sweden and Britain, violence is muted; in some, like Ireland, it is an historical aspect of politics; in still others, violence is associated with change—for example, the American civil rights movement and, earlier, the struggle for unionization. Where revolutionary violence can recruit beyond a single narrow generation and achieve the toleration of the many, this is an indication that societal efficiency is eroding or that inchoate nationalism exists. If there is to be order and justice, either there must be reform and accommodation to the dissent on which the gunmen feed or a form of devolution from the center must be achieved.

In Italy the South Tyrol threat was vastly eased after nearly ten years of intricate negotiation, and other separatist threats have been met by constitutional structures and special rewards. On the other hand the Italian system remains sufficiently inequitable and ineffective that the ideological gunmen feed on genuine dissent. Without change there may be more dissent and more gunmen. In France there has been little recognition that some form of regionalism may be necessary to ease the growth of separatism in Brittany, Corsica, and perhaps elsewhere. Centralism is a basic principle of the French Republic; and the need to shift, and shift rapidly and extensively, is not yet accepted in Paris. The Italians knew they had a nationality problem; the French have until recently barely noticed their Basques and Bretons. The worst of all prospects is to have a nationality problem without a solution. The British, by insisting that the Protestant majority in the six-county state carved out of Catholic Ireland in 1921 has a right to remain within the United Kingdom, have assured the continuing alienation of the minority. London cannot reimpose order with justice by democratic means, cannot establish a moderate provincial government, and so far has not seriously considered withdrawal. The United States, too, has a nationality problem without immediate solution, in that those who seek independence for Puerto Rico represent only a tiny majority of the island's population. To concede independence would be undemocratic.

The most serious internal problem for a democratic society, efficient or not, remains the submerged nation. Devolution is a difficult and trying procedure, federalism an uncertain answer, and novel forms often essential; but the alternative is the nationalist gunman, murdering for an invisible nation. His colleague who husbands only a radical ideology, a sense of grievance, and a messianic vision of the future can in a fragile and untried democracy be a terminal threat; but mostly the Angry Brigade or the Weathermen or the Second of June "guerrillas" disappear into jail, an early grave, or more conventional pursuits. Thus, although transnational and international terrorists are irrelevant to democratic societies, fanatics seeking a stage, not all terrorists can simply be ignored. The nationalists possess a real legitimacy and the ideologues indicate societal fragilities.

Consequently, the primary responsibility of those at the center is to gauge the nature of the terrorist threat. To transform democratic norms in order to hunt down and punish the Jackals is a policy that has little to recommend it. Jackals are not deterred by law nor by the failures of their predecessors. To make terrorism more difficult for the gunmen—as Sweden did through its emergency legislation focused on the emigré community—is not an unreasonable approach, especially since it eases the general anguish over the country's being used as an arena for alien battles. Although little can be done to assure an end to the transnational and international terrorists, a mix of techniques, technology, and international cooperation can do much to make their operations more difficult.

For democratic countries the rise of indigenous gunmen who may have a respectable tradition or real grievances requires an effective and carefully considered response. Recognizing them is not always easy. In the United States, in the turmoil of the 1960s, the Black Panthers wanted to carry guns, not shoot them, a fact the Chicago police did not understand; the SDS wanted to talk about revolution, not make one; the Black Liberation Army needed a rationale for their deadly fantasies, but were not actually interested in politics. The symbolic bombings of the 1970s in the United States are not serious revolutionary activity—not terrorism, but propaganda by the dissident, of little danger to anyone but the bomber and the odd innocent.

But there are also real problems—and those problems demand a response. The United Kingdom, with the exception of Northern Ireland, one of the most tranquil of societies, had a *real* problem in Ulster that was not simply "radicals" using the fashionable rhetoric of the day, symbol and display, nor alien gunmen seeking a stage. For nearly fifty years after the establishment of the Northern Ireland assembly at Stormont, "Ireland" disappeared from British politics. The majority could maintain order in their state without recourse to London. When in 1969 order could not be maintained and the institutionalized injustice of five decades was revealed, London had an Irish problem again. It was a problem no one in British politics wanted and was bedded in an area almost no one understood. First, the Labour Party and then the Tories delayed, moving lethargically, avoiding severe political surgery and permitting the British Army to maintain order in such a way that by 1971 Britain faced across the Irish Sea an insurrection with goals beyond easy accommodation. The delay in London was crucial, although no one, even with hindsight, can be assured that swift remedial action would have, in fact, prevented the slide to chaos. By 1972 the problem was that there was no solution—this, even in hindsight, did not seem to be the case in 1969–1970. Today, in spite of the violence in the Spanish Basque country and the growing alienation of the Bretons, swift, effective devolution of power from the center may reduce the unreconcilable to a frantic few and may establish new political forms on the model of the South Tyrol negotiations, rather than new battlegrounds, like Ulster.

Thus, revolutionary violence must be analyzed with great

care in order to weigh the reality of the challenge. The threat may be rhetorical, in which case it need be met only with sufficient innovation to ease public anxiety. The threat may be an indication of real grievance—an unpopular war, a restrictive university system, a rising rate of inflation—and here the grievance must be the first priority, not the protestors or proclaimed revolutionaries. Or the threat may be real, as in the case of Basque nationalism, and here the options are repression or accommodation.

The analysis, the sorting out into categories, the balancing of challenges, is not all that simple. The British let Irish matters drift for a great many good reasons. The French may be right in discounting the seriousness of the Bretons or Corsicans. And the Italians or the Spanish, in the midst of political crisis and change, may find an exacting examination of the threat of neo-Fascism too difficult.

Consequently, although perhaps the most important admonition—"Know Thy Terrorist"—is easy to offer, it is not always so easy to follow. If the evidence has been that to minimize the threat is the better policy, the reverse—an exaggerated concern with the gunmen—has been the preferred policy. After all the power and legitimacy of the state have been overtly challenged—and there have been sufficient examples from the past to indicate that some of those revolutionary challenges have brought down democratic societies. Democratic states rarely need admonitions to seek out revolutionary enemies and hone the tools of repression. That is the normal response. But then, politics is not an exact science.

If governments are going to respond to terrorism within a context of political opportunity and priority, and as a result of special perceptions and avowed or unavowed prejudice—as they are now and always have—grandiose, overarching strategic prescriptions are bound to be largely inapplicable. Even arrangements at the center are more form than substance. In the United States, the Cabinet Committee to Combat Terrorism met only a few times after 1972. The Working Group of the Committee represents twenty-six departments, but the representation is often by officials with little if any bureaucratic leverage. The group is a coordinating body, not a control center. The Department of State's Office for Combatting Terrorism has five officials and two secretaries. Ambassador Heck, State's representative on the Cabinet Committee, in 1976–1977, and one more in a string of appointments to the post, left in 1977 to become Ambassador to Nepal, some indication of State's priorities. There is thus really no command-and-control. The decision, if there is a major political component to an incident or a number of potential victims, may end up in the Oval Office or in the province of the National Security Council. The whole field of prior planning, too, leaves much to be desired. In December 1976, for example, the General Service Agency and the Federal Preparedness Agency estimated that thirty-four federal agencies would be seriously involved in any peacetime nuclear emergency. And the American experience is hardly unique.

No Western democracy can really be expected in the future to adopt a coherent national policy toward a terrorist threat, however defined. When President Nixon announced that the United States could not give in to terrorist blackmail, he expressed a posture and an aspiration, not a policy. Of course, the United States could and did give in—and in certain circumstances undoubtedly it should have. While it is a splendid idea to know of your enemy, in the United States after ten years and millions of words and scores of meetings, there is still no agreement on the nature and intentions of that enemy. What can be done is to recommend techniques and tactics with a low political component and to commend attitudes. The former always has greater charm to the official mind, for it suggests action. Yet attitudes are obviously more important and can, improbable as it seems, often be grafted onto traditional procedures—not, for example, the growing police acceptance of hostage-bargaining as a law enforcement tool.

Escalation: Less Is More

Neither police nor politicians like to wait on events when swift action can be seen to pay immediate benefits. The public, too, wants aberrant crime punished forthwith. At times, and under certain circumstances, illicit, violent challenge has become ritualized and the response can be gradually escalated under tolerant public eyes. In Paris, where riot is ritual, the French government has a series of security police formations, each more military than the last. These can be employed as matters become more serious. If political violence can be met with conventional forces and procedures, as the British have done in England with the IRA, then not only is the public comforted but the normal institutions of the state remain untouched. If the militia or the army is needed, then the general rule is "swiftly in and swiftly out" so that the police may again take over. It was in fact the tactical presence of the British Army in Ulster month after month with no strategic political initiative from London that led to insurgency. The British Army acting like an army all but created an insurrection simply by following conventional military procedures. London erred both on a tactical plane, by eschewing the principle of less is more, and on a strategic plane, by refusing to recognize the reality of grievance.

Beyond the restraint of law enforcement agencies, the law itself need not be greatly expanded. If scarecrow law frightens crows, it need not be more heavily clothed—and no scarecrow frightens Jackals. The death penalty will not deter the dedicated but only ensure martyrs, nor will martial law prevent the urban guerrilla who feeds on injustice. To prevent IRA leaders from speaking on television, as Dublin has done, will not limit their appeal and may prevent the many from seeing first-hand those leaders' contradictions and frailties. True, the call for "no free speech for traitors" has appeal; but freedom of expression is crucial in an open society. Similarly, any emergency legislation, no matter how minor the restriction on liberty, must be approached with great caution. Coercion laws tend to remain on the books, and the definitions of "emergency" grow broader until the "enemy" comes to include the dissenter, the alien, and the eccentric. Yet, there *are* emergencies. Intimidation, extortion, and murder can close down the judicial system, cripple the police, and leave brute force as the last defense for liberty. So most states have some form of emergency legislation—a War Powers Act or provisions for martial law—and some, like Ireland, have a vast legislative arsenal that excludes suspected individuals from most privileges of the law.

Some police forces follow the most stringent guidelines and others are given the broadest powers of search, investigation, intelligence, arrest, and interrogation. In the United States new guidelines for the FBI in the post-Watergate era reduced the number of investigations into domestic security cases from 4,868 to 214. There was still sentiment that the FBI should be restricted solely to criminal investigations. There is, in fact, nowhere a rigorous study of whether emergency legislation and expanded police power is effective. Nothing about arguments for *more* emergency powers is certain except that the motives of the advocates must be parsed carefully; and while special powers may be needed, then again they may be urged for partisan purpose.

Although less may be more, it is often necessary that less appear to be more, that a response, even a moderated response, satisfy an easily outraged public by its high visibility and assumed effect. Thus some laws can be recommended not because they will work but because they will not and yet seem to. Measures of deterrence must at least *appear* to work. The concept of sky marshals on American airplanes and the publicity surrounding their presence was a far more effective deterrent than the marshals' actual physical presence. The actual prevent-profile that weeded out the psychopaths largely remains a trade secret, so much so that even some of those in policy positions remain convinced that the decline in hijacking was related to the death penalty for air piracy, or the marshals, or a mix of factors. Ideally, then, deterrents work to dissuade the potential terrorist and to assure his potential victim. The great target airline, Israel's El Al, is often the choice of discerning travelers because the threat has produced the world's most stringent precautions. El Al passenger filtering is long, detailed, and thorough; the air marshals shoot to kill; and even as Entebbe proved there is hope of rescue after everything else has gone wrong. Every El Al rider is a symbol of defiance, and each feels in his or her way that every safe passage is a small victory over the men of violence.

The major premise of less is more relates to the irreversibility of a state response to challenge. In hostage-negotiation it is impossible to return to conversation after gunfire. In riots the moves up from bull-horns, through batons, to CS gas, rubber bullets, and at last real bullets are seldom reversed. The pressures to escalate are always there. The Dutch cabinet simply ran out of patience with the South Moluccans. Talk was leading nowhere. What was not recognized is that the stalemate was a positive step—no one was being killed. But after the marine commandos went in and broke the stalemate, people *were* killed. No real thought was given to in-between measures. The Dutch official arguments against continued talk seemed unconvincing, certainly to the relatives of the dead.

In matters of emergency legislation, the same caveat applies. The laws stay on the books. Law enforcement people become used to acting with special powers. The recent American experience of dismantling federal police and intelligence capacities is unusual and may be a passing matter, certainly if the necessary and proper investigation into subversive and criminal activity is so hampered that a backlash results. So, as always, there are no clear lines, only the need for a questioning attitude coupled with pragmatism—a bias that less is more but nothing is inadequate.

Retaliation: Beyond the Law, the Court of Last Resort

The temptation of regular armies to adopt the irregular tactics of their opponents—even when those tactics are criminal, illicit, and unsavory—is a perpetual temptation. The irregulars seem to have absolute freedom to act as they will, without the restraints of the law or society, and the regulars do not want to deny themselves that same freedom. In almost all cases the temptation must be resisted, because to act in a manner perceived as illicit erodes the legitimacy of the state. Although Israel sees the deeds of the Wrath of God within a war context, few others accept the premise. Few other democratic states have truly contemplated such acts for long—the Americans in response to Castro Cuba or the French in response to Algerian arms smuggling—and never as a long-term national policy.

It is not merely authorizing agents to kill for the state that must be avoided, but also the idea of vengeance at justice's expense. If the British had reintroduced the death penalty after the no-warning car-bombs, it would have been seen, cetainly in Ireland, as the Crown's revenge against specific individual Irishmen: others in England might murder for gain or in the pursuit of fantasies, but Irish men killing for political purpose must be killed in turn. The most important aspect of the law in fragile democratic societies is that, in theory, justice is blind. Consequently, calls for the death penalty are almost always a result of specific anguish rather than general analysis. If the state can avoid killing, except in self-defense and as a last resort, it is a stronger state. If harsh penalties are imposed for acts of political violence in a moment of hysteria, there will be those who see the penalty as retaliation beyond the law—and the separation of law from justice assures that order will be in jeopardy. This does not mean that an eye cannot be taken for an eye if this is just, legal, and traditional, nor that society must suffer endlessly the assaults, especially the unpunished assaults, of the fanatics. It simply suggests that the greatest defense of liberty within a just, open, and plural society is recourse to disinterested law, law that protects the many rather than only specifically punishes the few.

Conclusion

If counsel is of no use after the fact, it might be wise for those in authority to consider in advance the dangers of the next decade—perhaps a time of terror, but certain to be one of turmoil. Some want—and need—a terrorist threat. Some truly believe that Carlos and his friends are a terminal threat to Western civilization. Almost all recognize that the public does *not* have a prepared mind, will insist on a visible response, and at best can only slowly be persuaded that less is often more. Ultimately what is needed is a flexible pragmatism, a confidence in the moral basis of the present society, a willingness to contemplate accommodation and concession, especially to real grievance, and maintenace of the close linkage between law and justice rather than law and order. While free democratic societies are rare and have proven fragile, they too have their strengths. It is in fact these strengths that politicians and the public most readily want to erode when faced with the gunmen. The right to dissent, to advocate treason, to pursue in public strange fantasies should be protected and cherished, not restricted in the name of order. Maniacs have the right to plead; the media, to cover in intimate detail the massacre of the moment.

What is so distressing is that many of those responsible for the massacres of the moment do not represent legitimate dissent in an open society but are rather driven by fantasies or struggle for alien dreams, as was the case of the five Croatian hijackers. There is almost no evidence that anyone in Crotia wants independence. The Croatians are a threat to open societies only insofar as these societies feel impelled to close down, even close down marginally as did the Swedes. The fact is that anytime, anywhere, in authoritarian as well as democratic states, a determined band can probably seize an airplane filled with hostages. The filtering of passengers works only to a degree. All technology and techniques work only to a degree. Murphy's law ("If anything can go wrong, it will.") is everywhere operative, as it was at the bomb disposal pit in the Bronx. And the complex technology of the present means that a few with iron pots can cause immeasurable complications for the many. Those who strike at the fringe areas of control in the international system, as did the Croatians, find room to manuever, each threatened government eager to avoid the poison parcel, unable to control events at a distance, uncertain of the course of events. And in Paris when the end came, there was no reasoned response, no patient siege, no experts hastily summoned, but rather the inevitable outraged indignation made manifest in the ultimatum of the head of state. Still, the Croatians did surrender, were sent back to the United States, and were tried and convicted under existing law. Their conviction in no way persuaded three other Croatians that the risks were too great and, in June 1977, they seized the Yugoslavian embassy in New York and shot and wounded an employee.

There is no way, then, to protect open societies at all times from the violent men. There is no way to transform the media so that it can protect itself from capture by terrorist choreographers. There is no way to make every nuclear facility absolutely safe or every airport perfectly secure. And so far there seems no way to persuade the outraged public that repeated assaults on these vulnerabilities are inevitable, that the most effective response is for the public to prepare its mind and compose its soul. Calls for vengeance, even under laws' cloak, are understandable, ill-advised, and to little effect.

For the government, there is no hope for an apolitical strategy or for effective strategic planning or even for detailed prior preparations. That is not the way the political animal operates in Western society. Much, however, can be done that is not counterproductive, that may even have political advantage. The techniques and tactics of deterrence and defense, high technologies, and the refined crafts of the social scientists can be deployed. New developments may put nuclear facilities beyond terrorist reach or raise effective hurdles to hijacking. Coupled with efforts to restrict sanctuary, to draft reasonable and workable conventions and treaties, to wage a diplomatic campaign against terror, the very existence of new laws will have a calming effect. And the law, of course, does work—the Croatians are in prison.

There are as well tactics of response, effective attitudes toward provocation; but most important, there should be a growing realization that many policies, good, bad, and indifferent, recommended to those in authority come without buttressing evidence but replete with special pleading. Terror seems to bring out the worst in all those involved, not only the hijackers but also the editorial writers and politicians and professors who grasp any opportunity to urge their own vision on a troubled world. Thus the opponents of nuclear power have through wishful thinking summoned up nuclear terrorists, have already inspired symbolic bombs and guerrilla raids, have created the genii they professed to fear. Politicians, even of the Left, in Ireland and Italy, have enacted laws of the moment that they later may feel a necessity to repudiate or, as in Ireland, carry down with them in electoral defeat. And academic careers founded on a terrorist threat or an international conspiracy or the laws of air piracy can hardly go forward to further publication and all-expense-paid conferences if "terrorism" evaporates.

Beware of those offering solutions to terrorism. There are no solutions in open societies. There *are* advisable attitudes, incremental protection from the violent, and the assurance, as yet without takers, that the greatest danger from the enemy is that the enemy is us. The terrorists cannot bomb down an open society, but an act of parliament can close one. The terrorists can, however, represent real grievance, as they do in Ulster and the Basque country and Italy. But there, too, absolute repression would be incompatible with an open society. In Belfast and Genoa and San Sebastian, the gunmen are real too, their banners patent, their causes logical—unlike the men with the iron pots or the Jackals who can be hired with a slogan. The latter must, however difficult, be treated as irrelevant to reality; the former must be recognized as danger signs requiring reform or the devolution of power, as well as repression.

At the end for those who will never discuss matters of policy with ministers or generals, visit the Oval Office on business, or receive the midnight call from an anguished airline official, the crucial advice remains the same. Terror in its manifold forms will remain with us. Sometimes such violence is significant as a real threat, but mostly it is not. The enemy is us. Indignation is expensive; outrage is dear. Make the best of a troubled world. Do not open bulky packages mailed from an unfamiliar address in Belfast. Avoid riding with controversial diplomats, applying for executive positions in troubled zones, or flying in planes that accept unfiltered passengers in the Rome or Athens terminals. Do not vacation in Uganda or lunch with Italian judges. The world is largely free of small pox and the plague, but not of hurricanes or terrorists. So, most of all, don't despair simply because we live in interesting times. Perilous as matters seem, if open, democratic societies in the West cannot protect the liberty of us all from a handful of gunmen, accommodate legitimate dissent, and repress the politics of atrocity under the law—if we cannot tolerate the exaggerated horror flashed on the evening news or the random bomb without recourse to the tyrannt's manual—then we do not deserve to be free.

BIBLIOGRAPHY

One of the sure and certain signs of a novel and popular topic in academic eyes is the emergence of bibliographical literature—in the fullness of time there may even be bibliographies of bibliographies, even for so seemingly a narrow subject as terrorism. In splendid isolation, of course, terrorism is not very interesting; but once the bounds are stretched to include psychological factors or revolutionary movements or the law of war, adequate coverage becomes extremely difficult. In the glory days of antiinsurgency in 1961, the Special Operations Research Office (SORO)—sponsored by American University in Washington, D.C., and funded by the United States Army—began to publish bibliographies, one of the major academic scissors-and-paste operations of all time, and one that implied the interested would have to spend their time reading lists of books rather than books themselves. Today, there are massive bibliographic materials on many phases of political violence. For example, there are special institutes and journals and collections of documents concerning the contemporary Palestinian problem, and no end in sight. On the other hand, for some terrorist movements, there is little readily available except exclusive proclamations and the records of the court.

A good, keen selective bibliography on *Violence, Internal War and Revolution* is that of Michael J. Kelly and Thomas H. Mitchell (Ottawa: Norman Paterson School of International Affairs, Carleton University, April 1976); its 496 items, including a special section on Canadian violence, give a reasonably representative selection rather than the massive, almost counterproductive, coverage of the SORO approach. A similar, more selective approach on a broader scale by William H. Overholt, *Revolution: A Bibliography* (Croton-on-Hudson, N.Y.: Hudson Institute, January 9, 1975), contains forty-three pages in what is a working document related to Overholt's study of revolution. A variety of scholarly treatments also have excellent and detailed bibliographies—see, for example, Ted Robert Gurr's *Why Men Rebel* (Princeton: Princeton University Press, 1970), which focuses on the nature of frustration-aggression theses in matters of political violence.

One of the first attempts to focus solely on terrorism was made by Lyn Felsenfeld and Brian Jenkins in *International Terrorism: An Annotated Bibliography* (Santa Monica, Calif.: RAND, September 1973), which listed a great deal of ephemeral journalism and little not published in English. This seems to have been a failing as well with many of the early American bibliographic efforts, especially by governmental agencies. The reverse is true with Roger Cosyns-Verhaegen, *Present-Day Terrorism: Bibliographical Section* (Wavre, Belgium: Centre D'Information et de Documentation de La. L.I.L., 1973), which lists 131 works, mostly in French. Fortunately for those concerned, all the quibbles no longer especially matter, since Edward F. Mikolus of the United States Central Intelligence Agency has produced a major and massive *Annotated Bibliography on Transnational and International Terrorism* (Washington, D.C.: C.I.A., December 1976). And—a sign of the times—it is unclassified. Despite the fact that Mikolus excludes works on American domestic problems and those not related to international terrorism, his bibliography contains 1,277 items (again, mostly in English) in 225 pages. While some of the areas reveal largely speculative works or, in the case of the geographical areas, the tip of the iceberg, Mikolus has performed a significant scholarly exercise of considerable use—for those who need the best on bombs or the latest in disaster response.

For those who want not lists but the best works on the subject, there is a profusion of riches. On guerrillas in general, the best recent treatment is Walter Laqueur's *Guerrilla: A Historical and Critical Study* (Boston: Little, Brown, 1976) and the companion *The Guerrilla Reader: A Historical Anthology* (Philadelphia: Temple University Press, 1977). The final third of Laqueur's trilogy, *Terrorism*, is still to come. A superb anthology is *Revolutionary Guerrilla Warfare* (Chicago: Precedent, 1975), edited by Sam C. Sarkesian and containing work by both the practitioners and the professors. For those who want a first-hand approach to insurgency and rebellion in more detail (T. E. Lawrence aside), some of the basic works from one side of the barricades would be:

Menachem Begin, *The Revolt* (New York: Henry Schumer, 1951).
Regis Debray, *Revolution in the Revolution* (New York: Grove Press, 1967).
Vo Nguyen Giap, *People's War, People's Army* (New York: Praeger, 1967).
———, *Big Victory, Great Task* (New York: Praeger, 1968).
General George Grivas, *Guerrilla Warfare and EOKA's Struggle* (London: Longmans, 1964).
Che Guevara, *Guerrilla Warfare* (New York: Random House, 1965).
Ho Chi Minh, *On Revolution* (New York: Praeger 1967).
Mao Tse-tung, *Selected Military Writing* (Peking: 1963).
Abdul Haris Nasution, *Fundamentals of Guerrilla Warfare* (New York: Praeger, 1965).
Troung Chinh, *Primer for Revolt* (New York: Praeger, 1963).

Works from the other side of the barricades include:

Brigadier Frank Kitson, *Low-Intensity Operations* (Harrisburg, Penna.: Stackpole, 1971).
Sir Robert Thompson, *Defeating Communist Insurgency* (London: Chatto and Windus, 1966).
Colonel Napoleon D. Valeriano and Lieutenant-Colonel Charles T. R. Bohannan, *Counterguerrilla Operations: The Philippine Experience* (New York: Praeger, 1966).

Naturally, there are a great many more memoirs, scholarly investigations, anthologies, journal articles, and various forms of hagiography from when today's terrorist was yesterday's guerrilla. Two important earlier works are Brian Crozier's *The Rebel: A Study of Post-War Insurrections* (Boston: Beacon Press, 1960), and Roland Gaucher's *Les Terroristes* (Paris: Albin Michel, 1965).

Once the guerrilla came in from the rural backlands and became an urban terrorist, and once the hijackers created a fashionable revolutionary technique, there was a rush to explain the phenomena. Thus we were given general works on the terrorist, often less than felicitously stitched together:

Charles Atala and Ethel Groffier, *Terrorisme et Guerrilla: La Revolte Armee Devant Les Nations* (Ottawa: Dossiers Interlex, Les Editions Lemeac, 1973).

J. Bowyer Bell, *Transnational Terror* (Stanford, Calif., and Washington, D.C.: Hoover Institution and American Enterprise, 1975).

Jacques Bergier, *La Troisieme Guerre Mondial Est Commencee* (Paris: Albin Michel, 1976).

Anthony Burton, *Urban Terrorism* (New York: Free Press, 1975).

Richard Clutterbuck, *Living with Terrorism* (London: Faber, 1975).

———, *Protest and the Urban Guerrilla* (London: Abelard-Schuman, 1973).

John Gellner, *Bayonets in the Streets: Urban Guerrillas at Home and Abroad* (Don Mills, Ontario: Collier-Macmillan, 1974).

Frederick J. Hacker, *Crusaders, Criminals, Crazies: Terror and Terrorism in Our Time* (New York: W. W. Norton, 1976).

Edward Hyams, *Terrorists and Terrorism* (London: J. M. Dent, 1975).

I. Matekolo, *Les Dessous du Terrorisme International* (Paris: Julliard, 1973).

Gerald McKnight, *The Mind of the Terrorist* (London: Michael Joseph, 1974).

Robert Moss, *Urban Guerrillas: The New Face of Political Violence* (London: Temple Smith, 1971).

Albert Parry, *Terrorism from Robespierre to Arafat* (New York: Vanguard, 1976).

David C. Rapoport, *Assassination and Terrorism* (Toronto: Canadian Broadcasting Company, 1971).

Lester A. Sobel, ed., *Political Terrorism* (New York: Facts on File, 1975).

Francis M. Watson, *Political Terrorism: The Threat and Response* (Washington, D.C.: Robert B. Luce, 1976).

Paul Wilkinson, *Political Terrorism* (London: Macmillan, 1974).

There are as well a considerable number of articles treating terrorism as a general problem rather than dealing with a single facet—far too many articles to cite here. More effective than a random sampling of articles is a brief list representing a variety of approaches or case studies collected in anthologies:

Yonah Alexander, ed., *International Terrorism: National, Regional and Global Perspectives* (New York: Praeger, 1976).

Carol Edler Baumann, ed., *International Terrorism* (Milwaukee: Institute of World Affairs, University of Wisconsin, 1974).

David Carlton and Carlo Schaerf, eds., *International Terrorism and World Security* (London: Croom Helm, 1975).

James F. Kirkham, Sheldon E. Levy, and William J. Crotty, eds., *Assassination and Political Violence,* Report to the National Commission on the Causes and Prevention of Violence (New York: Bantam, 1970).

Jay Mallin, ed., *Terror and the Urban Guerrilla,* (Coral Gables, Fla.: University of Miami Press, 1971).

J. Niezing, ed., *Urban Guerrilla: Studies on the Theory, Strategy, and Practice of Political Violence in Modern Societies* (Rotterdam: Rotterdam University Press, 1974).

Bard E. O'Neill, D. J. Alberts, and Stephen Rossetti, eds., *Political Violence and Insurgency* (Arvada, Col.: Phoenix Press, 1974).

There are as well three other collections currently being prepared:

Yonah Alexander and Seymour Maxwell Finger, eds., *Terrorism: Interdisciplinary Perspectives* (New York: John Jay, forthcoming, 1977).

Marius Livington, ed., *Terrorism in the Contemporary World* (Westport, Conn.: Greenwood Press, forthcoming, 1977).

Michael Stohl, ed., *The Politics of Terror: A Reader in Theory and Practice* (New York: Marcel Dekker, forthcoming, 1977).

There are also the records of conferences and seminars; whole issues of journals dedicated to the subject (*The Skeptic,* January–February 1976), or at least large selections (*Orbis,* vol. 19, no. 4, Winter 1976); and long sections of testimony before Congressional committees.

Out of all this welter of material there are individuals of various persuasions and disciplines who have produced a considerable literature or cornered a theoretical approach or a special niche. One of the most prolific—and one who has not limited himself to matters of political violence—is Brian Crozier, whose Institute for the Study of Conflict has produced a long and most useful series of monographs (see, for example, Paul Wilkinson's "Terrorism Versus Liberal Democracy—The Problems of Response," *Conflict Studies,* 67, January 1976) and the *Annual of Power and Conflict.* Crozier has written in a variety of journals—as early as 1959 in *Nation,* on "The Anatomy of Terrorism." He is the core of the Union Jack School consisting of Clutterbuck, Moss, and others.

An equally typical but American approach has been that of Brian Jenkins of RAND, except that he has eschewed the trendy tools of social science, often chosen by his colleagues, with a whole variety of widely scattered scripts intended mainly for policy use; the articles sum up his evolving conclusions: "International Terrorism: A Balance Sheet," (*Survival*); "International Terrorism: A New Mode of Conflict" (California Arms Control and Foreign Policy Seminar); "Terrorism Works—Sometimes" (RAND); "High Technology Terrorism and Surrogate War: The Impact of New Technology on Low-Level Violence" (RAND); "Combatting International Terrorism: The Role of Congress" (RAND); "Will Terrorists Go Nuclear?" (California Arms Control and Foreign Policy Seminar); and "International Terrorism: A New Mode of Conflict" (*International Terrorism and World Security*). Jenkins has testified before a variety of Congressional committees, presented a paper on future trends in terror at a State Department conference, and is a consultant to the Nuclear Regulatory Commission—but you can seldom find his work on the book shelves or in scholarly journals.

Other American scholars have approached the problem with special techniques (Phillip A. Karber, "Urban Terrorism: Baseline Data and a Conceptual Framework," *Social Science Quarterly*, December 1971, pp. 521–533), or educated particular audiences (Jay Mallin, "Terrorism as a Political Weapon," *Air University Review*, vol. 22, July–August 1972, pp. 45–52), or speculated on demand (Irving Louis Horowitz, *Political Terrorism and Personal Deviance*, United States Department of State, XR/RNAS-21). A great many are concerned with a single aspect of the problem and appear in appropriate academic forums (Alona E. Evans, "Aircraft Hijacking: What Is Being Done," *American Journal of International Law*, vol. 67, October 1973, pp. 641–671; or Bernard Feld, "The Menace of Fission Power Economy," *Science and Public Affairs*, vol. 30, April 1974, pp. 32–34; or Leon Romaniecki, "The Soviet Union and International Terrorism," *Soviet Studies*, vol. 26, 1974, pp. 417–440). A few, like Professor Horowitz, when asked to speculate, stayed to write and lecture repeatedly on the problems posed by terrorism, especially to civil liberties. Others, like Martha Crenshaw Hutchinson, arrived in the field without prior planning—in this case, with an elegant dissertation on the Algerian FLN and revolutionary terrorism—and moved on to such matters as terrorism as a policy issue or the problems of nuclear terrorism. Those who were concerned with the American violence and turmoil of the 1960s often reworked their data, as have Ted Gurr and Ivo K. and Rosalind L. Feierabend. Some come into the terrorist academic arena for a single display, and others can be found nowhere else. There is something for nearly every reader, often, it seems, by nearly every specialist.

Concerning the actual acts of violence, nowhere have so many written from such diverse perspectives as in the matter of hijacking—the archetypal contemporary act of terrorism. Just a selection of the more popular genre, excluding most of the scholarly works, adds up rapidly: James A. Arey, *The Sky Pirates* (New York: Scribner's, 1972); Peter Clune, *An Anatomy of Skyjacking* (London: Abelard-Schuman, 1973); David G. Hubbard, *The Skyjacker: His Flights of Fantasy* (New York: Collier, 1973). From the victim's view, there is Uri Oren, *99 Days in Damascus: The Story of Professor Shlomo Samueloff and the Hijack of TWA Flight 848* (London: Weidenfeld and Nicolson, 1970); from the hijacker's, Leila Khaled, *My People Shall Live: The Autobiography of a Revolutionary* (London; Hodder and Stoughton, 1973). And there are mounds of governmental reports, psychological studies, and accounts of various aspects, especially on the Palestinian fedayeen; for example, Peter Snow and David Phillips, *The Arab Hijack* (New York: Ballantine Books, 1971).

There is a veritable library concerning assassination—in general, in particular, in theory, and in practice. Those interested in bombs can learn how to make them by reading *The Anarchist's Cookbook* or what to do about them in Thomas G. Brodie's *Bombs and Bombings: A Handbook to Detection, Disposal and Investigation for Police and Fire Departments* (Springfield, Ill.: Charles C. Thomas). Most of those who have been involved in hostage-bargaining have written on their method or delivered papers or been interviewed at length—most also are convinced that they alone discovered the technique. Thus, if one is concerned about a single incident, there is often a massive tome available, such as Serge Groussand's *The Blood of Israel, The Massacre of Israeli Athletes—The Olympics 1972* (New York: Morrow, 1975); if the concern is a category of horror, there will surely be a script: Murray Clark Havens, Carl Leiden, and Karl M. Schmitt, *The Politics of Assassination* (Englewood Cliffs, N.J.: Prentice-Hall, 1970); or if the concern is a special movement, most have a historian, some tame, some scathingly critical like Jillian Becker in *Hitler's Children: The Story of the Baader-Meinhof Terrorist Gang* (Philadelphia: Lippincott, 1977).

It is the problem of gathering sources for specific terrorist movements that turns back even the most determined bibliographer. The material available on Fatah in English alone is mind-boggling. And yet if there has not yet been a definitive survey, the sources are raw ore—court cases, party manifestos, published interviews, and the first cut of journalism. If the struggle is not over, attempts to publish—on the Provisional IRA or the Basque ETA, for example—are soon overcome by events. There are all kinds of splendid historical studies, but if the common reader on terrorism seeks immediacy—information on the career of Carlos or the intentions of the Japanese Red Army, not the relevance of the Macedonia Black Hand or the Spanish anarchists—he or she is often left with potted history or quick and dirty journalism. To some extent Crozier's conflict studies by various hands can keep one up to date on many such matters; but the ultimate and definitive studies on *Brigate Rosse* in Italy or the ERP in Argentina—to mention just two examples—must wait on events.

Beyond the studies on the practitioners of violence, there has obviously also been a massive literature produced on the appropriate response to the gunmen and bombers: *cf.* "The Common Wisdom." And, except for the specialist, even a sampling is a daunting task. Mikolus in his bibliography has fifteen pages on Potential Nuclear Threats and three more on Disaster Response (indicating we are more threatened than safeguarded). And there is a vast literature in varying tongues and degrees of classification that Mikolus did not or could not list. In the BDM study "Analysis of the Terrorist Threat to the Commercial Nuclear Industry" (Vienna, Va.: September 30, 1975) there are dense columns of government documents, and no one really knows how many internal studies have circulated, been lost or filed, forgotten or revised. And this is only the American government.

Abroad, in other democracies, journalists and scholars and bureaucrats, too, churn out pages on nuclear terrorism or the problem of the media or the most advantageous legal approach to terrorist acts.

What was needed was a basic book. And in sum my former graduate student, while embarked on study of such matters at MIT, suggested that all anyone need know about terrorism could be gained from seeing the film *Battle of Algiers* ("The Wit and Wisdom of Ernest Evans," unpublished), which is quite probably true; however, in a democratic country one needs to know more about the appropriate response to terrorism. But, for the moment at least, there still seems to be too much to be described, much less easily digested.

INDEX